Rafik Davidovich Grigoryan

Know thyself: synergy and antagonism of cells

Rafik Davidovich Grigoryan

Know thyself: synergy and antagonism of cells

ScienciaScripts

Imprint

Cover image: www.ingimage.com

This book is a translation from the original published under ISBN 978-620-2-39638-7.

Publisher:
Sciencia Scripts
is a trademark of
Dodo Books Indian Ocean Ltd. and OmniScriptum S.R.L publishing group

120 High Road, East Finchley, London, N2 9ED, United Kingdom
Str. Armeneasca 28/1, office 1, Chisinau MD-2012, Republic of Moldova, Europe
Managing Directors: Ieva Konstantinova, Victoria Ursu
info@omniscriptum.com

Printed at: see last page
ISBN: 978-620-8-51095-4

Contents

Preface

Dear reader!

I am pleased to recommend a popular science book by a well-known theoretical physiologist, Doctor of Biological Sciences Rafik Davidovich Grigoryan.

In trying to understand the complex, we build on what is known and use analogies. The human body is a very complex object. We know that it is made up of different types of cells. The author uses the properties of cells to explain the properties of the organism. This is the main novelty of the book.

The complex biological patterns of the human body are presented so lucidly that they can be understood by a person far from biology. Instead of the standard description of health and disease issues with reference to anatomy, the author has chosen a different strategy. He views health problems as an inevitable concomitant result of long biological evolution. The key stages of this evolution are highlighted: the emergence of the nucleusless cell, the cell with a nucleus, the efficient way of its energy supply, the emergence of cells of different types and, finally, the multicellular organism as a fundamentally new form of life. Ontogenesis, the life of an individual, is briefly described.

The author's method of presenting modern knowledge of biology through dialogues with an artificial intelligence expert makes the reader a participant of live discussion of the material, which, in my opinion, will help inquisitive readers with a technical background to assimilate the knowledge presented.

The author not only uses facts and established scientific truths, but also analyses some controversial problems of modern biology and medicine. Despite the fact that I was familiar with the main scientific publications of the author (the list of monographs is given on the last page of the book), some interpretations of facts given in the book turned out to be new for me as well.

Physiologists have previously proposed concepts of holistic functioning of the organism, but they are incompatible. In the recommended book, the author has reduced the extremes into one alternative concept. Overall, the book is very timely: for the first time human physiology is presented as a consequence of the synergy and antagonism of specialised cells. It is the systemic understanding of the organism that is so lacking in modern medicine of narrow specialisations.

The book is designed for the curious person who wants to know the basics and inner workings of his body. I believe the reader will not only deepen his erudition, but will also get real pleasure from reading this book.

Corresponding Member of the National Academy of Sciences of Ukraine,
Doctor of medical sciences, Professor V.F. Sagach

Foreword by the author

The writer Vladimir Korolenko (1853-1921) coined a phrase in his essay "Paradox": "Man is born for happiness, like a bird for flight". In the essay, a man born without hands writes these words with his foot. Thanks to Maxim Gorky, this phrase entered the public consciousness and became part of the humanistic perception of the world. We like to think so, but it would be more correct to say that man, like any living being, is born to be healthy.

It is true that most people are healthy for most of their lives. Nevertheless, it is rare not to have experienced temporary states of ailment or illness. With the help of a doctor, and often without a doctor's intervention, health is restored. Why? The short answer to this question is that the body has normalisation mechanisms. Rarely is anyone able to answer the question of what these mechanisms are and how they came to be. The answers to exactly these questions are contained in the book you have opened.

The book is structured in the form of dialogues between a physiologist and an engineer. I have tried to convey to the reader the basic ideas about life in general in an accessible and entertaining form. The basics of ontogenesis, the rules of coexistence of our cells, the fundamental role that energy plays in the work of natural mechanisms of cell life are considered. It also answers the question of why, as we age, maintaining the optimal state of cells and their populations (colonies) becomes more difficult, and health becomes unstable, fragile, and steadily weakening.

There is another humanistic problem, which is the drama of intelligent life.

Neither the ancient Greeks, who created theatre and proposed its main genres, nor later playwrights have revealed the source of the greatest drama of the universe - human life. And this drama consists not only and not so much in the fact that every human being is mortal, but in the fact that in order to maintain his life he has to eat other living beings.

The realisation of this drama did not come at once. At different stages of human civilisation and in different closed communities, food prohibitions appeared asynchronously but consistently. However, cannibalism is still practised in some places today. But I would like to focus on the inherent problem of intelligent life. This problem arises in the mind of everyone who, having sufficient intellect, tries to abstract himself from the daily worries of providing decent conditions of existence and getting pleasures from its separate manifestations, asks himself a question: is the very principle of reasonable life based on animal instincts fatal?

All previous attempts of mankind to find a satisfactory answer to this question or its variations only gave birth to religion and morality, which, over time undergoing their changes, directly or indirectly regulate the personal and social life of each of us. Both morality and religion in their own way explained man, defined his place and role in nature, tried to humanise man-beast. But these attempts in no way reflected the true reasons for the combination in man of his morality with clearly animal instincts. The roots of this dilemma remained invisible until the theory of evolution appeared. But for the bulk of society, evolution remained a speculative concept with no direct relevance to man. Only when it became clear that the bodies of all macroscopic organisms, including us, consist of a huge number of microscopic specialised cells, and that all life on Earth is located on the same genetic tree, did scientists begin to look for the root causes of health and disease in genes, in the properties of these cells and the patterns of their interaction. Science has already passed a certain part of the road to disclosure of individual rules of cells, but there is still a lot of work to be done to fully understand them.

In the book offered to you, I have tried to explain how our bodies have evolved evolutionarily, what basic cellular properties determine our physiology, and why we sometimes get sick.

I would like to dwell on the worldview aspect of the book.

Since science and the scientific worldview itself is evolving, I will start from afar.

With the invention of the telescope, humanity's observational knowledge of celestial bodies increased dramatically. The telescope made it possible to detect objects in the near space that the limited sensitivity of the human eye prevented us from seeing. Thanks to the fact that modern telescopes have made visible also radiation beyond the visible range of electromagnetic waves, many mysteries of the physics of stars have been revealed. The recent extension of telescopes into outer space has been a new observational breakthrough for astrophysicists. But no matter how advanced the observational tools, cosmologists do not limit themselves to measurements: they use these data to create speculative and mathematical models of the evolution of the cosmos. It is thanks to these models that the Big Bang theory emerged, and science was able to explain not only the processes of star formation, but also the origin of chemical elements of the Mendeleev table.

The invention of the microscope made it possible to see objects inaccessible to our vision because of their small size. The microscope is an indispensable tool for biologists. They saw cells and individual details of their internal structure with their own eyes. The resolving power of modern electron microscopes allows to distinguish not only biological polymers and macromolecules. Molecules and atoms are already available for observation.

It would seem that, armed with such technologies of observation of the structural organisation of living things, biologists and medics should have long ago found out the rules and laws determining the coexistence of specialised cells of the human body. However, an organism is a very complex system of specialised cells. One microscope alone, however perfect it may be, cannot understand this system. Therefore, there is still no knowledge with the help of which it would be possible to carry out accurate early diagnosis of complex diseases, their effective prevention, as well as reliable treatment.

What's the matter?

In my opinion, the point is that there is no proper comprehension of the accumulated data. In other words, there is no adequate theory of life of a multicellular organism.

A theory has to be created. But a theory cannot be built only on scattered facts, even if there are many of them. The value of a theory that is limited to reproducing only known observations is small. Such a theory would have only didactic value.

There is a need for such a theory, with the help of which it would be possible to predict new facts, to replace expensive and often unsafe empirical studies with calculations on quantitative mathematical models. The good news is that modern computers are an effective tool for the researcher. It is the solution of biological problems with the help of modelling and simulation experiments that I have been engaged in for more than four decades. Many scientific articles and six monographs have been published.

For a long time I had been hatching plans to write a popular book. Chance helped.

A good old friend of mine, who has been working on artificial intelligence in the USA and Germany for a long time, recently approached me with an interesting request. He said that his boss, also a specialist in artificial intelligence, had called him in a few months ago and set him a completely new, unexpected task - to create a mathematical model to investigate the relationship between human intelligence and its physiological mechanisms. He also said that the boss wanted to start a new long-term research project, for the financial support of which there seemed to be sponsors. But they require a presentation on how the human body functions in a comprehensible way. They are also interested in whether it is possible to take a fresh look at health problems and offer fundamentally new technologies for the prevention, diagnosis and treatment of non-trivial diseases, as well as slowing down the ageing process.

I realised from my friend's words that all his attempts to understand the principles of the human body by reading existing scientific publications had not led to anything good. Rather on the contrary, they only made him furious: the publications either explained everything very popularly with emphasis on the obvious anatomy of the organs, or were full of terms, neither the essence nor the relationship between which he could not grasp. In general, a dead end....

Towards the end, my friend said the following:

- Firstly, I know that you model the human circulation. I even remembered our recent conversations about you offering some new insights into human physiology and publishing several monographs. Well, who better to turn to for help than you. Step into my shoes. Help me.

I caught the flattery, of course, but I still wondered....

I myself had the desire and motives to write about my science in a popular way. Firstly, I became convinced that the scientific basis of modern medicine is rather shaky and often based on myths. Secondly, after the collapse of the USSR, the former materialistic worldview has eroded. All sorcerers and shamans blossomed in society, who for a lot of money are ready to send any unfortunate sick person safely to the other world. Their financial possibilities allow them to buy time for advertising on TV screens and in newspapers. In a short period of time they have managed to confuse people so much with their medieval magical rites and obscurantist theories that even among my academic colleagues, some of whom are biologists by education, there are many who, as they say, "have gone mad".

This book is not about religion, and I have no desire to attack anyone's deep religious beliefs. However, I believe that there is no one but scientists to oppose the drift of science into the dark maze of religion and magic. This drift is evident - I have seen the Bible and icons in the office of a doctor-academician, I know other academicians, formerly convinced materialists, but now baptised and believing in the existence of magicians, otherworldly life, heaven and hell, devils and angels. I am convinced that these inventions of our uneducated ancestors have no place in academia.

Sometimes I entered into dialogue with some of today's believing scientists, trying to understand the reasons for this drift of consciousness and the logic of the new views. However, the general picture was depressing: very soon the interlocutor seemed to cease to be a participant in the dialogue, and a cold and aloof glint appeared in his eyes. The facts and arguments I presented bounced away without reaching the depths of the interlocutor's consciousness. Perhaps it was the age of these people. In old age, the brain is fragmented, trying not to lose the thread of the narrative, it clings to individual words and concepts that sound familiar. But such a brain is no longer able to perceive a multifaceted unified picture of the world. Therefore, in the heads of such people the boundaries between the possible and the impossible are blurred. The sieve of logic gapes with big holes: everything is possible.... More than once, seeing such a depressing picture, I was convinced that facts and logic cannot break through the concrete wall of faith!

One often hears that science shies away from uncomfortable phenomena. Science does not have to have ready-made answers to all questions. It does not undertake to interpret rumours and so-called dubious facts. In my opinion, the most important thing in science is that it as a method of cognition, allows you to separate the class of possible phenomena from the impossible. The progressive movement of science gives hope that the phenomena, unexplained today, will eventually find their logical explanation. As a scientist, I see the only way to prevent the spread of obscurantism is through enlightenment. It should be directed at those whose minds still retain at least pockets of logical thinking, and whose seeds of knowledge have a chance to sprout. I believe that the backbone of my readers are inquisitive people, first of all, those who have received a mathematical or technical education and are inwardly opposed to obscurantism.

My previous attempts to write a popular book about how our bodies ensure our health and the origins of non-communicable diseases proved difficult for the layman. This time it seemed to me that by responding to a friend's request, I might be able to solve my own problem. So I thought...

While talking on Skype, my friend said that he was taking a two-week holiday soon, and that he would come and meet me live and talk to me. I was happy and promised to give a definite answer by the time he arrived.

From that point on, I was in contemplation.....

On the one hand, I wanted to help my friend. On the other hand, I doubted whether I could explain my views on life to him in such a way that they would become his worldview. After all, only in this case he would be able to transfer these views to a third person. The matter was aggravated by the fact that many details of the organism's functioning were still unknown to me. I had only intuitively reached some large-block logical generalisations. Although the new empirical data that I am gleaning from various scientific publications reinforce my confidence in the validity of my theory, its experimental proof will require a long time and new experimental techniques. Nevertheless, I decided to take the case and soon informed my friend about

it. I had a condition.
I asked that he prepare to listen to me for a week, maybe a little more. I also asked that everything be recorded on a tape recorder so that before the next conversation he could listen to it again and mark the parts he didn't quite understand.
He agreed. We had thirteen conversations. In the following I outline an adapted copy of the audio material.
A friend has taken it upon himself to translate a written text into English. I already know that this text was studied by my friend's supervisor. He was satisfied with it. This gives me reason to believe that finally, in a first approximation, my task of writing a popular book about my ideas about the workings of the body has been successfully accomplished. I have therefore ventured to publish it, adding only a preface.
I hope that interested readers, at least those close to engineering thinking, will also draw useful information from this book.
I am grateful to Tatiana Ludovik for helpful advice and editing of the manuscript. I am responsible for any inaccuracies.
Please send feedback and comments to the following e-mail address: rgrygoryan@gmail.com -
12.06.2018 г.

CHAPTER 1

Talk one

Introduction. The body as a community of specialised cells

- In order to understand the workings of such a complex natural mechanism as the human organism, one needs a correct worldview. Its cornerstone is the acceptance of the fact of evolution. Moreover, not only biological evolution, but also the evolution of the universe.

Scholastics formulated the chicken and egg problem. Within the framework of simple logic, this problem remained unsolvable for a long time. However, the solution came from the other side - with Charles Darwin's theory of the origin of species. It turned out that neither chicken nor egg emerged in a completed form, but is the result of evolution of previous ways of reproduction.

Unfortunately, many people associate the word "evolution" only with Darwin's theory. Those who do not accept this theory are in fact opponents not only of this biological theory, but also of evolution in general. Meanwhile, everything that surrounds us is the result of evolution and continues to evolve. Those who have studied philosophy are familiar with Hegel's statement that internal contradictions are the source of movement and development. Translating this statement into the language of natural sciences, it is easy to see that statics can be only temporary, when the vector sum of all acting forces is equal to zero. However, processes have inertia. Therefore, nothing can stay at the point of equilibrium for any length of time.

With the discovery of the law of conservation of energy, it became clear that energy can have different forms. Their mutual transformations cannot be eternal if there is no external source to replenish the expended portions of energy. This is the essence of all cyclic processes. They are known in physics and chemistry. Fundamental biochemical transformations are based on them. Even the so familiar cyclic act of heart contraction would not be possible if the resting potential and excitability were not restored in myocardial cells during diastole (the relaxation phase of the heart muscle). This restorative process against the natural concentration and electrochemical gradients between the cytoplasm and the extracellular fluid is maintained because the cell uses special macroergies synthesised beforehand. Life, from its origins to our appearance, has utilised various available sources of external energy in order to energetically support internal cyclic processes. But it would be wrong to focus solely on biological evolution. Otherwise, the illusion arises that nature created the mechanism of evolution for the purpose of creating us

- thinking beings.
Intelligence is not the only or the most important achievement of evolution. Life forms have long and successfully existed and reproduced without intelligence. The dinosaurs that ruled the planet for about 150 million years were not gifted with intelligence. But they had other advantages in the struggle for existence. Crocodile shows an even more perfect model of the animal: different species of crocodiles were before the era of dinosaurs, co-existed with them and survived the catastrophe of the end of the Jurassic period, which wiped out all large specimens of dinosaurs. It is true that a different branch of dinosaurs, which we know today as birds, also survived this catastrophe and adorn the modern fauna with their colourful plumage and forms. Moreover, intellectless creatures continue to live alongside us and other animals with different levels of intelligence. Evolution has equipped them with sufficient mechanisms for survival in the face of changes in their environment. Many species of protozoa have survived more than one biological catastrophe.
It has been established that in all known major geological shifts, the more complexly organised species disappeared first. There is no reason to believe that we are an exception. To date, our intelligence has not made us safe from either external catastrophic threats or internal threats associated with our animal nature. Morality and religion only partially mitigate civilisational contradictions. But even morality and religion evolve, although this evolution has different internal driving forces.
Today we know that even the chemical elements of the Mendeleev table are the product of a long evolution of matter, its original energetic form. Although physicists' knowledge of this original substance is only hypothetical, science has nevertheless established the time when it all began. And it began about 13.8 billion years ago with the expansion of a very dense and mysterious form of matter (pramatter). It is sometimes thought of as a special form of plasma.
Science has established that as space expanded and the plasma cooled, gluons and quarks gradually formed, later forming the basis of electrons and protons. The interaction of protons with electrons gave rise to the hydrogen atom. Many hydrogen atoms were formed in space. Gravity gradually thickened the hydrogen clouds and, at a certain density, the local temperature increased so much that stars and their systems united by gravity began to form.
In a star, expansion forces act against gravity. The latter is a by-product of the temperature, which rises during nuclear fusion reactions. As long as these two forces are in balance, the star is stable. Nuclear fusion reactions of

chemical elements go all the way up to the formation of an iron atom. The more massive atoms in the Mendeleev table require more intense energy to synthesise. It comes in the next phase of a star's life cycle. When the entire initial supply of hydrogen is exhausted, the above balance of opposing forces is broken. Gravity takes over, the star is unrestrainedly compressed, and as the atoms of matter come closer together, its temperature and pressure rise rapidly. The star explodes, briefly becoming a supernova star. In this process, a number of chemical elements heavier than iron are formed. (The back cover of the book depicts the major stages in the evolution of our universe and life on Earth. The collage is made of adapted pictures taken from the Internet).

From the exploded star, these chemical elements are scattered in space and mixed with the existing hydrogen atoms. The density of matter in such clouds is not evenly distributed. Gravity is stronger in places of the highest density. Clots of matter are formed there, and the process of matter concentration in them continues with increasing speed. Any external shock (for example, the shock wave of another supernova explosion) initiates the formation of a new star. This is already a second-generation star. Around it, planets are formed from a part of atomic matter.

So, there is a planet, there are chemical elements. At a certain convergence of these elements, chemistry begins - individual elements, combining, form chemical compounds.

Let me remind you that chemistry is the generalisation of electrons in the outer orbitals of the constituent atoms of molecules. Any chemical fusion reaction requires energy.

I would like to draw your attention to the fact that according to the second law of thermodynamics, without energy supply, all structures eventually disintegrate. This is true of all biological macromolecules. They are constantly breaking down into smaller components. Any biological structure consisting of bricks - macromolecules - can exist for a long time only if macromolecules are re-synthesised at a rate adequate to the rate of their disintegration. This fundamental position of biochemistry as a sword of Damocles threatens the existence of each of our cells and organism as a whole.

- Wait a minute. Because the second law of thermodynamics applies to all of space. How did life get round that law?

- Life is a local phenomenon. The energy required for the construction of complex biological macromolecules is drawn from the nearest local environment.

- How did she end up there?

- That's a good question! In the case of terrestrial life, the primary source of energy is the Sun. But science believes that the transition of chemistry into biology, i.e. the emergence of life, is somehow connected with the accumulation of external energy in the form of chemical bonds. Their breakdown supplies the local intracellular environment with the energy necessary and sufficient to perform all forms of work against the second law of thermodynamics. In particular, for the formation of more complex cell structures from biochemical components.

I note that science does not yet know how the first cell was formed. And we will return to the question of energy many times. The question of energy and driving forces is key not only for understanding biodynamics. It remains key also in physics, when we study the dynamics of mechanical, electromagnetic objects. I am convinced that the role of energy in the organism has not yet been fully elucidated.....

Be that as it may, an important milestone in biological evolution was the fact that some early representative of life (believed to be an archaic organism called a cyanobacterium, or blue-green bacterium) was the first cell to master the chemical reaction of converting the energy of sunlight into special macromolecules - carbohydrates. These stored molecules then serve as an intracellular energy source.

Modern biology cannot speculate evidentially about whether this foremother cell of life on Earth formed from the chemical elements of our planet, or whether its birthplace is not Earth. Cyanobacteria has no nucleus, but it has a number of remarkable features that have played a key role in bringing you and me into existence. First of all, it should be noted that cyanobacteria produce a by-product during their metabolism: oxygen. For a long time on the early Earth, oxygen was a poison to microorganisms.

Cyanobacteria do not have mitochondria. A mitochondrion is an organelle in our cells that, among other ingredients, uses oxygen to synthesise special molecules - energy accumulators. They are also called macroergases. The most powerful of these is the molecule adenosine triphosphoric acid (ATP). We will talk more about it later. Here we will only note that ATP is a universal supplier of energy in all our cells, as well as in all *eukaryotes* (cells with a nucleus) and multicellular organisms.

- Where did the cage come from?

- There is an ongoing debate in biology about this. It is known that all life forms are based on the cell. How it came to be is not known. One hypothesis believes that it arose under some favourable conditions on Earth in an aqueous environment rich in macromolecules. There's also the panspermia hypothesis. According to it, the age of the Earth is not sufficient for chemistry to give rise to biology. The panspermia hypothesis suggests that life in the form of a single-celled organism was brought to Earth from outer space with objects that are much older than Earth.

Be that as it may, it is known that on our planet many species of unicellular organisms coexist with many species of multicellular organisms. According to modern ideas based on genetics, all biological objects on Earth belong to a single tree of life. And life itself, as a unique phenomenon that has been sustained for about four billion years, owes an important ability to reproduce itself by division (*mitosis).*

Here I want to interrupt my summary of the evolution of life and make an important note about what a cell is.

A cell is a certain isolated area of space, within which a specific, self-sustaining process of synthesis of specialised organelles takes place, the coordinated functioning of which makes it possible to draw energy and building chemicals from the outside and process them to provide three basic functions of life: 1) growth; 2) reproduction by mitosis (division); 3) response to external stimuli. The totality of chemical reactions within the cell is summarised by the term *metabolism.*

You should know that self-sustaining cyclic processes occur in physics and chemistry as well.

- Yes, I once read about Belousov's reaction. At least, I remember that he observed how a liquid spontaneously cyclically changes colour. I even read that Alan Turing, whom we all know as a computer theorist, wrote a special book in which he explained patterns on animal bodies by random processes similar to Belousov's reaction. Finally, I also read that the process of crystal formation is considered by some experts as a precursor of life.

- I'm glad you're so aware. But I'd like to get back to the cage.

The cell is structurally separated from the outside space by the cell membrane. It is permeable to small chemical elements, water and ions, but impermeable to relatively large chemical molecules. It has special gate mechanisms made of proteins. They have two positions: open and closed. In the closed position of the membrane gate, the inner part (protoplasm with organelles floating in it) contains everything necessary to maintain the metabolism of the cell.

Generally, there are two kinds of cells - nucleusless (*prokaryotes*) and with a nucleus (*eukaryotes*). Our cells belong to eukaryotes. But some cells (for example, blood cells) in the adult state are deprived of a nucleus.

The nucleus of a cell contains the hereditary apparatus - DNA. In humans,

DNA consists of about 25,000 genes. All DNA is fragmented into unequal parts in the form of chromosomes. We have 23 pairs of them (i.e. 46 chromosomes).
The number of chromosomes has nothing to do with the complexity of the organisation of a biological object. To confirm this idea, I will give just a few examples. The greatest number of chromosomes is in the invertebrate marine animal radiolaria - 1600, in the fern plant there are 1200 of them. The smallest - in one species of ants (two in the female, one in the male). The number of chromosomes in a pea (it was on it that Gregor Mendel discovered the laws of heredity) is 12, in rice - 54, in a cat - 38, in a dog - 78, in a cow - 120. Our closest relatives, chimpanzees and gorillas, have 48, and macaques have 42.
At the DNA level, the genetic differences between different people are 10 times smaller than the differences between humans and chimpanzees. Genetic differences between men and women are due to the presence of the Y-chromosome in men. It contains only 86 genes (the X chromosome contains from one thousand to one and a half thousand genes).
No offence to us men, it has been said that women have 1520% more grey matter (neurons) in their brains. Although the brain mass of women is generally smaller than that of men, the intelligence score, as determined in psychological tests, is not lower. It's just that a woman's brain contains more active elements in a smaller volume. But men have two other advantages: more white matter and intracerebral fluid. White matter, the inter-neuronal cables. - allow for better distribution of tasks between different parts of the brain. Intracerebral fluid, contained in the ventricles of the brain, cushions blows to the head. Our ancestors needed this quality more than you and me!
Thus, the number of chromosomes indicates the evolutionary path taken rather than reflecting the organisational complexity of a creature. This statement is all the more valuable in view of the fact that most genes of species, including ours, are not active, i.e. do not encode proteins. Discussions regarding the evolutionary role of silent genes continue unabated.
- What are mitochondrial diseases, and why are they only transmitted through the maternal line?
- I was going to talk about mitochondria some other time. But since the question has been asked, I'll answer it.
Mitochondria are organelles that contain their own genes. In the male sex cell, the spermatozoon, there are very few mitochondria, and they only serve to give the spermatozoon the energy it needs to swim to the egg by twisting

its tail-motor. But the egg has quite a lot of mitochondria. In the process of formation of all specialised cells of the body from a fertilised egg (*zygote*), as well as in all subsequent acts of mitosis, mitochondria inherited from the egg are pre-doubled and transferred to the daughter cell. This is what concerns the essence of mitochondrial gene inheritance.

Before turning to mitochondrial diseases, let me mention a curious genetic study conducted after the human genome was decoded at the beginning of the 21st century. The study involved people of both sexes of all races and ethnic groups. It turned out that all currently living mankind - are the descendants of one woman (she was called mitochondrial Eve), who lived in central Africa about 200 thousand years before us. This also implies that those mitochondrial diseases, which appear only in certain ethnic groups, are the result of late mutations.

- Funny, was there a Y chromosomal Adam?

- Yes, he was. According to the latest data, he lived about 338,000 years before us.

- It got even funnier: genetic Eve and Adam lived at different times.

- You have to understand it this way: all men alive today have a version of the Y chromosome, the oldest analogue of which was found in the genes of a man who lived about 338,000 years before us. Of course, the sons of the so-called mitochondrial Eve got their male chromosome from a man living in her time.

Now for the diseases.

Mitochondrial diseases are caused by the presence of mutations in the mitochondria. There are about five dozen such diseases known. All of them are directly or indirectly caused by the fact that mutant genes of mitochondria impair their main function - aerobic synthesis of ATP molecules.

- Got it.

- Let's keep talking about our cells.

It is important for us to remember another figure: there are 220 cell types in the human body. A cell type is its specialisation. For example, everyone knows that we have nerve cells (neurons), muscle cells *(myocytes),* liver *cells (hepatocytes*), several types of skin cells, bone cells and others. All these cells are formed from the initial - fertilised egg of the mother organism. In the adult human body, the number of cells is estimated in the tens of trillions. In other words, each cell type forms a population (sometimes the word *"colony"* is used*) of* sister cells.

In addition to our own cells, which share a common genome, there are about ten times as many foreign microorganisms in the body and on its surface. In fact, the relationship between our cells and non-human cells is such that the human body cannot survive for any length of time without the majority of foreign microbes. This is primarily because the vast majority of microbes (about 1.5kg) are found in the gut and, by digesting the food we eat, convert it into *nutrients* (nutrients) that are good for us. Other microorganisms inhabiting the skin of our body protect us from the attacks of harmful bacteria. In short, we are the product of a symbiosis between our genome and the thousands of times greater total genome of our protectors. Understanding this symbiosis is a prerequisite for understanding the mechanisms of health and disease.

- Wow! And my doctor parents taught me to fight germs. I'm still afraid of germs. So hygiene is wrong?

- The answer is yes and no. Disturbances in the diversity of intestinal microflora lead to inefficient digestion and inadequate supply of vitamins of all kinds. It is the intestinal microbes that produce most vitamins. Science cannot yet accurately trace all the effects of digestive disorders, but some causal chains have been established. It is believed that the sharp increase in the number of patients with various forms of allergies may be due to excessive personal hygiene. This does not mean, of course, that hygiene should be neglected. But there is no need to be overzealous either.

- Well, you scared the hell out of me.

- Okay, let's limit ourselves to that body of knowledge for today. I believe you've received enough information to think about. We'll continue this time tomorrow.

- Thank you. It's a deal.

CHAPTER 2

Talk two

Ontogeny

- So, did you learn what you heard yesterday?

- I think so. Although the topic was so vast that I would never have brought these phenomena down to some sort of unified framework before. But I guess you have some kind of vision of the commonality and philosophical connection between stars and life. It is good that we are not talking about fashionable nowadays nonsense of astrology and energy-information interactions. You know that as a person who graduated from physics and engineering institute, I can't stand this nonsense.

- Then let's move on. Today I want to enlighten you about ontogenesis.

- What's that?

- In general, this term denotes the individual development of an organism, i.e., the set of successive morphological, physiological, and biochemical transformations undergone by an organism from fertilisation to the end of life.

Human ontogeny begins with that happy moment when one of millions of spermatozoa (the smallest male cells), having overcome all the obstacles on the way to the ovum (the largest cell of the female organism), passed through its membrane, reached the nucleus and transferred its 23 chromosomes to it. By fusing the male chromosomes with the 23 unpaired maternal chromosomes, *a zygote* is formed. In the nucleus of the zygote, 23 pairs of chromosomes are formed. They contain almost all the information that will be present in every cell of the future organism.

- Why "almost"? Isn't all genetic information contained in the nucleus?

- Not all of it. There is a small part of DNA that is passed on to the next generation exclusively from the mother. This part of the DNA is contained in the mitochondria of the egg. The mitochondrion contains just over 37 genes. But even these genes make a difference. By the way, there are a number of hereditary diseases, carriers of which are precisely mitochondrial genes.

The word *"mitochondrion"* should be spoken with the utmost respect. Without mitochondria, there would be no you and me. It is hard to imagine multicellular life in any of its variations without them.

- What are they doing in the cage that makes their role so important?

Mitochondria are often referred to as the energy substations of the cell. The mitochondrion deserves this characterisation because it synthesises ATP molecules, the breakdown of which releases energy. We need it for many things, from keeping the cell alive to ensuring our thermoregulation.

- Why do you use the words *"mitochondrion"* and *"mitochondria"*? Are there a lot of them?

- Yes. Every cell has many mitochondria, from a few units to hundreds or more. The number of mitochondria in a cell correlates with the rate of energy expenditure. Mitochondria also vary in size.

We will talk more than once about mitochondria and the molecules they synthesise. Now I would like to talk about how mitochondria came to be.

The most common hypothesis is that mitochondria were once independent microbial organisms. Their distinctive feature was that, as a result of mutation, they learnt to use oxygen in the reactions of ATP synthesis. Before this, there was already one way of synthesising ATP from glucose in the cytoplasm of the cell. This method is called anaerobic (i.e., without oxygen) glycolysis. But its efficiency is low - only two ATP molecules from one glucose molecule. But mitochondria use the product of glycolysis - pyruvic acid - and in a chain of successive transformations with the participation of oxygen create another 32 ATP molecules.

- Whoa! The cage is 17 times more powerful?!

- Exactly! As a man with an engineering background, you must realise that such a leap in energy capability meant a lot. Energy is power, potential.

Now imagine that such a microbe, the ancestor of the mitochondrion, was swallowed by another cell. It did not digest it, but there was a clever exchange of genes between them. And part of the microbe's genes entered the nucleus of the cell that swallowed it, and about 40 genes remained to work only in this microbe-mitochondrion.

- Can I interrupt you for a moment? Does this mean that some genetically modified product that we eat could replicate the path of this mitochondrion and introduce its genome into our cells?

- Got it. It's a fairly common concern. The thing is, our digestive system doesn't pass large biological molecules like genes. It breaks them down into fragments that pose no threat.

Now back to oxygen. It is noteworthy that before the appearance of the mutated microbe, oxygen was a poison for all kinds of anaerobic cells. And a lot of oxygen had already been released into the atmosphere by blue-green organisms. And so, the mutation produced an organism for which oxygen turned from a poison into a useful ingredient for internal energy supply. Such a zigzag of evolution!

It is believed that it was this cell line that later gave rise to a new type of organisms - multicellular organisms. After all, for the formation of such a

community of cells and its survival in an aggressive environment, extra energy is a huge competitive advantage. Man is only one of hundreds of millions of such organisms.
- I get it. What you're saying is that evolution isn't stupid to refuse such an acquisition.
- I'm glad to hear that conclusion.
Let us return to ontogenesis. Let me note that in human ontogenesis there are two periods - perinatal (the embryo is formed in the womb) and postnatal (during the rest of life). Ontogeny is often contrasted with phylogeny - the slow changes an organism undergoes during the evolution of a species. Phylogenetic evolution can last hundreds of millions of years.
I'd like to enlighten you on some of the issues of early embryogenesis and review *gastrulation.*
- Why do I need to know that? Is it that important?
- I think so. You've heard of stem cells, haven't you?
- I know, I've heard of it. I've even read some. They called them stem cells because there is a certain analogy between a tree of cell specialisation and an ordinary tree. From a common trunk, branches branch off and grow. So it seems to be the case with our cells. But it seems to me that all this is somehow far from the general framework of our discussions. Correct me if I'm wrong.
- Not that you're so wrong, but I think what I'm about to inform you will serve to form a proper worldview.
- I'm sorry, but I forgot the word you said. Something like pot...
- By the way, not a bad sounding analogy: gastrulation - pot. Something from the kitchen. And we're talking about the kitchen of life, where the dish "embryo" is cooked. So be it, you will memorise, associating gastrulation with the words kitchen and pots.
I want to show you a picture that schematically depicts the stages of gastrulation (Fig.1).
Gastrulation is the process of formation of three layers of cells of different specialisation *(germ sheets).* Gastrulation in humans occurs in two stages. The first stage is by cell division and the second stage is by cell migration.
During the first stage, two germ sheets (ectoderm and entoderm), two provisory organs (amnion and yolk sac - not shown in Fig. 1) are formed. Immediately before the first stage begins, the formation of the following takes place
of the provisor organ, the precursor to the placenta.

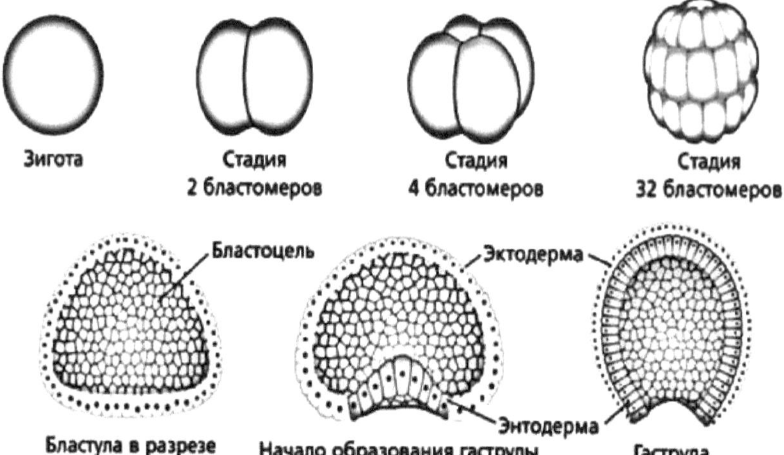

Figure 1. Gastrulation: stages of transformation of the zygote into precursor cells from which specialised human cells are formed.

- This is all, of course, interesting, if only because nature has solved such a complex problem without intelligence. But, in my opinion, when creating an artificial organism, there is no need to copy all natural stages.
- I in no way meant to say that everything will have to be repeated. You and I have a basic engineering background and we know that natural solutions are not always the best. I just wanted to show you that cell specialisation is a sequential process. So far we have looked at the simple initial steps. After all, next, these cells must form organs. I will say
you - some of them work only with a certain sequence of constructing a dozen varieties of cells. The places where they form may be different. But the cells then migrate and somehow find an assembly site. Embryologists have only observational information about these phenomena, but their mechanisms remain a mystery. Perhaps, from here we can bridge to the phenomena I was planning to inform you about. I will only note that later on this way I will load you with additional information about the structure of chromosomes.
Be that as it may, please remember one detail of the picture of Figure 1, namely, that the zygote produces new cells by the method of fragmentation. We will come back to this feature a little later.
We have already said that our body contains 220 types of cells. They are all descendants of a single fertilised egg and were formed during embryogenesis. Although every cell in the body carries the same genetic set, each specialised cell has its own part of functioning genes.

Moreover, you have probably heard or read that specialists are already able to reverse the process of cell specialisation, i.e. they can make a stem cell out of any specialised adult cell. After all, a stem cell is a cell at all the early stages of its specialisation, i.e. stem cells are of different kinds, the main thing is that it should not be a finally differentiated cell.

Stem cells are divided into several groups:

Totipotent stem cells. They can differentiate into cells of embryonic and extraembryonic tissues organised as three-dimensional connected structures (tissues, organs, organ systems, organism). Cells formed during the first few cycles of zygote division are also totipotent.

Pluripotent stem cells are descendants of totipotent stem cells and can give rise to almost all tissues and organs. From these stem cells, three germ sheets develop: ectoderm, mesoderm and entoderm.

Multitypotent stem cells give rise to cells of different tissues, but the diversity of their types is limited to the limits of a single germinal sheet. The ectoderm gives rise to the nervous system, the sensory organs, the anterior and posterior sections of the intestinal tube, and the skin epithelium. From the mesoderm are formed cartilage and bone skeleton, blood vessels, kidneys and muscles. The human entoderm forms the intestinal mucosa, as well as the liver, pancreas and lungs.

There is also a distinction between *oligopotent* cells (can differentiate only into some similar cell types) and unipotent cells (immature cells, which, strictly speaking, are no longer stem cells, as they can produce only one cell type). They are capable of multiple self-reproduction, which makes them a long-term source of cells of one particular cell type and distinguishes them from non-stem cells. However, their ability to self-reproduce is limited to a certain number of divisions, which also distinguishes them from true stem cells.

- I'm going to have a hard time remembering these terms, but my understanding is that as you get further away from the beginning of the differentiation process (i.e., from the trunk), the potential to form groups of specialised cells decreases. Am I right?

- Yes, that's the way it is. Today, there are already clinics that use stem cell technology for various purposes. I would like to draw your attention to such a peculiarity of ontogenesis that our organs are formed from one of the three embryonic sheets. Roughly speaking, some organs are more similar in the characteristics of their cells than others. This feature is important for understanding physiological reactions of related organs to a certain group of endogenous or exogenous biologically active substances.

- Are you saying, I'm exaggerating, that one drug can affect one group of organs and another can affect another group of organs?
- Yeah, that's about right.
But I want to focus your attention on one seemingly strange peculiarity of embryogenesis. At least, I myself was very surprised when I first read about it. It's about the exact mechanisms that turn a non-specialised basic cell - i.e. a fertilised egg - into a specialised stem cell. I used to think that there was some strict DNA programme that tells each cell what it should become. It turns out it happens in a very different way.....
Are you saying there's no genetic predetermination?
- Exactly.
- Then why is DNA called a programme?
- Good question! I can explain it. I think there are two reasons for this. The first is thinking by analogy. It concerns a lot of borrowing. There is a drift of the original concept, strictly defined within one science (in our case, computer science) into other fields of science. The second is from ignorance of the essence. We always try to explain an unfamiliar phenomenon with the help of a familiar one, relying on signs of similarity. Only with time, when the details of the phenomenon become clear, we look for a more appropriate word for it.
- Although I said it in a florid way, I get your train of thought and I think I agree with you.
Nevertheless, the thought has not yet left me that there is no need for me to know such subtleties. Perhaps I will realise it later. With this hope I am ready to listen, tell me.
- Telling you.
I have already compared the sperm and the ovum in terms of size and said that the ovum is the largest cell in a woman's body. As for absolute figures, the diameter of the ovum is about 110-180 μm. It can be seen even without a microscope. Most human cells are a thousandth of the volume or a hundredth of the diameter of an ovum.
Due to the size of the fertilised egg, in the initial stages of embryo development, all changes are reduced only to the splitting of the original material into two equal parts. Thus, without breaks for growth (and this phase in the normal cell division cycle is quite long and energy-consuming), the fertilised ovum can divide into two cells, then into four, eight, sixteen, thirty-two and further without external material support. Everything necessary is available in the ovum.
This type of embryo fragmentation without consuming additional energy to

maintain cell division is very important because the embryo does not yet have any special connection to the maternal organism from which nutrition could be supplied. This will only come later, when the required communication structure is in place.

Here we should mention: in the process of fragmentation, all molecules contained in the cytoplasm are almost evenly distributed among the new daughter cells. But each of the cells will be viable only if it receives all 46 chromosomes from the mother cell.

- But there are only 46 of them in a fertilised egg.

- Yes. That's why DNA doubling (replication) occurs before each egg fragmentation.

I want to make an important insertion here, concerning the method used by nature in building a new organism within another.

Embryologists and other biologists note the striking difference between the natural way of creating a new organism and the technologies developed by man to design something artificial. Everything artificial requires a creator who has a blueprint of the object to be created. Moreover, at the intermediate stages of assembly, the object does not function. Its functioning becomes possible only after all elements of the blueprint are created and assembled into a whole. For example, even a table lamp does not shine until all its parts are assembled in a certain way and it is connected to the electrical network.

The case of a living organism is completely different: it gradually creates itself. This means that the structure assembled at a given step must immediately function in such a way as to ensure further self-assembly.

- Wait! But I have two fundamental questions. One, what is the power source that keeps it functioning? The second is where does the information about what to do next come from?

- These questions didn't just arise in your head. Biologists have asked them, too.

The answer to the first question was much simpler than the second. At the initial stages of embryo development, the egg is the source of energy, and then everything needed comes from the mother's body through the placenta.

It was long believed that the step-by-step information for embryo self-assembly was read from DNA. Turns out that's not the case.

The fact is that at all successive stages of embryo fragmentation, until each cell reaches the approximate size of the future specialised cells of the body, all newly created cells are genetically identical.

- Then the question is natural: without instructions from DNA, how does each cell know which cell it should specialise into - to become a nerve cell or a muscle cell, a bone cell, or a cell of the kidneys, liver, other organs and tissues?

- The point is that genetic identity does not mean complete identity of the conditions of life of different embryonic cells. It turned out that the specialisation of a cell is determined by the place it occupies in the embryo. It is one thing when it is located in the central region of the embryo and surrounded by similar cells, and quite a different thing if it happens to be on the outer side of a colony of cells.

In the latter case, the cell is acted upon from the outside by mechanical forces generated by the elastic properties of the outer membrane of the ovum, and from the inside by mechanical counter forces of sister cells. These two forces do not exactly balance each other out. It is this asymmetry of mechanical forces that is crucial for some genes in the genome to be activated and others not.

- And what does the word "activated" mean in this context?

- Active genes take part in protein synthesis. Passive genes do not manifest themselves in this case. But it is known that under a number of physical and chemical conditions, a gene can change its state from passive to active or vice versa. Therefore, the answer to your question is as follows: a previously silent gene starts to participate in protein synthesis.

To understand the relationship between mechanical forces and the activation of specific parts of the genome, we need additional information about the most important and well-known molecule of all organisms - DNA. Therefore, we will temporarily interrupt the description of embryogenesis and delve a little deeper into genetics.

The similarities between children and parents have been known for a long time. It was intuitive that something was inherited by children. But what and how?

The experience of cultivation of wild plants and domestication of different animal species led to certain rules of selection. It was realised that both parents play their role in shaping the traits of the offspring.

As early as the early 19th century, the French botanist Jean Baptiste Lamarck, wishing to explain the fact that organisms are well adapted to living conditions, formulated the idea that useful traits acquired by parents during their lifetime are inherited in subsequent generations. Charles Darwin's The Origin of Species by Means of Natural Selection, published in 1859, argued for an alternative idea. Note that Darwin knew almost nothing about patterns of inheritance. His supporters and followers tried to find explanations for Lamarck's observations from the standpoint of Darwin's theory. For this purpose, it was necessary to find out what the laws of inheritance are and what structures are the carriers of heredity.

The first laws of inheritance were not identified and formulated until 1865 by the Austrian botanist, Augustinian monk Gregor Mendel. He did this by observing how the external traits of peas are inherited in generations. He was lucky in this respect because, as I mentioned a little earlier, because of the small number of chromosomes (Mendel knew nothing about them!) peas have relatively simple rules of inheritance. Unfortunately, the results of Mendel's scientific endeavours remained almost unknown for a long time. It was only 35 years later that Mendel's laws of inheritance were rediscovered.

The German zoologist August Weismann initially adhered to Lamarckism. But Darwin's theory of evolution of species seemed more convincing to him, so he thought long and hard about the problem of transmission of hereditary traits to offspring and came to the conclusion that there must be natural carriers of these traits. Weismann first showed the general biological significance of mitosis and the fundamental role of the chromosomal apparatus in cell division. He proved that there must exist units of heredity and called them *ids*. In fact, he was referring to genes. Only much later, when the American biologist E. Wilson confirmed Weismann's observations about the activity of chromosomes, it became obvious that Weismann was right.

By this time, genetics had been taken up by Thomas Morgan, an established research biologist. Morgan confirmed that chromosomes were the material carrier of heredity. Morgan was awarded the Nobel Prize in Physiology and Medicine in 1933 for his discovery of the role of chromosomes in heredity. Being an honest scientist, Morgan not only found Mendel's work, but also recognised his priority in genetics.

Genetics has become a booming experimental science. It's rare not to have heard of Drosophila. These flies have become one of the favourite subjects of geneticists for three reasons. Firstly, Drosophila reproduce quickly. This quality is important if a researcher wants to see the effects of his manipulations in a series of animal generations. Secondly, Drosophila have only four chromosomes. Finally, keeping Drosophila is inexpensive. They breed successfully in the presence of ripe fruit.

Thanks to drosophila flies, not only the rules of inheritance of traits in subsequent generations were established, but also it became clear that there is a material carrier of heredity. And it is concentrated in the chromosomes of the cell nucleus. His search continued for a long time. Among other candidates for the role of a carrier of hereditary information, one curious molecule deoxyribonucleic acid was among the first candidates. Aka the

well-known DNA. But how exactly information could be transmitted from parents to offspring remained a mystery.
The first significant result was obtained after it was established that the DNA molecule is a double helix of RNA molecules. Moreover, the junctions between a pair of RNA molecules can be of one of four types. In 1953, this led James Watson and Francis Crick, who saw an X-ray image of a fragment of a DNA molecule, to a hypothesis explaining the procedure of transferring hereditary information in the process of reproduction of DNA molecules. They believed that at the first stage the molecules serving as connecting bridges between RNA pairs are broken and two RNA molecules are formed. Since a clear pairing correspondence was found between these two molecules, each of them became a matrix on which the missing pair of linking molecules could be assembled. In other words, the process of copying the original DNA molecule took place and two DNA molecules were produced instead of one. It is now known that this idea was not quite correct. In fact, RNA is formed during a process called transcription, that is, the synthesis of RNA on the DNA matrix, carried out by special enzymes - RNA polymerases. The matrix RNAs (mRNAs) then take part in a process called translation. Translation is the synthesis of protein on the mRNA matrix with the participation of ribosomes. This ensures that during mitosis (the division of a mother cell into two daughter cells), each new cell receives a complete set of instructions on how to make the next daughter cell.
It is known that the hypothesis of Watson and Crick was soon confirmed. Together with Maurice Wilkins, they received the Nobel Prize for their discovery in 1962. But, as it often happens in science, the process of inheritance of genetic traits turned out to be more complicated than it was imagined according to the double helix model of Watson and Crick.
Recall that DNA is present in every cell, but this molecule is fragmented into several parts of unequal length and is concentrated in chromosomes. The number of chromosomes in each species of organisms is different. As mentioned above, in humans this number is 46 (23 paired chromosomes). In each cell division (mitosis), the daughter cell receives a complete set of chromosomes.
Meiosis, the formation of sex cells, is also noteworthy. Biologists agree that it is thanks to sexual reproduction, in which the genes of both parents are shuffled, that the evolution of species has accelerated dramatically and species diversity has increased many times over. The main difference between meiosis and mitosis is that only unpaired chromosomes are formed. Therefore, a sex cell cannot divide on its own. To do so, it must fuse with

another sex cell of its own species to form a complete combination of paired chromosomes.

Proteins are the basic building and functional links in the structures and processes of life support. They consist of amino acids. The latter are synthesised within each cell according to instructions written in DNA. To read these instructions, you need access to them.

The genetic code is a way of encoding the sequence of amino acid residues in proteins using a sequence of nucleotides in a nucleic acid. There are four nucleic acids: adenine, guanine,
thymine and cytosine. Nucleotides are complex
nucleoside and phosphoric acid compounds.

In the introduction it has already been said that not all genes of the human genome are active, i.e. capable of expression. The fact is that DNA is not the only structure of chromosomes. To illustrate the place of DNA in a chromosome, a picture similar to Fig. 2 is usually used.

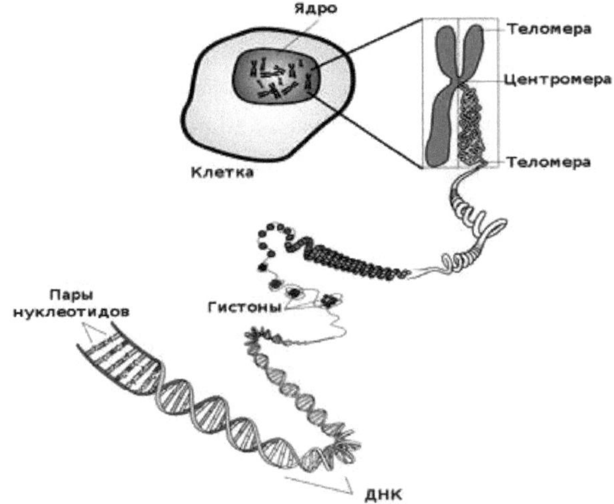

Figure 2. Arrangement of the chromosome and DNA.

Each DNA molecule is sheathed with histones - special nuclear proteins. Histones fulfil two main functions: a) they package DNA strands; b) they participate in epigenctic regulation of nuclear processes of transcription, replication and repair. I won't go into the details. As it can be seen from the picture in Fig. 2, in the

In the four-digit genetic code, the "letters of heredity" (*adenine, guanine, thymine* and *cytosine*) are located between the constituent RNA strands of DNA. It is important to know that out of the above four "letters of heredity", pair bonds between RNA strands are formed only in this way: adenine-thymine; guanine-cytosine *(principle complementarity*). It is due to this that each RNA strand serves as a matrix for DNA molecule repair.

- And why did it take a long time to decode the human genome?

- Not only time, but also money. I will not and cannot dwell on all the problems of this international project. I will only note that "words" of "letters" can have arbitrary length, and there are no gaps between "words" in the genome. By the moment of the beginning of work on this project the potential number of human genes was estimated as more than hundred thousand. This number was based on the then known number of proteins and on the concept of "one gene - one protein". As it turned out, this concept is wrong. The number of known proteins already exceeds 200 thousand, and many genes, often located in different chromosomes, take part in the coding of each protein. But there is another intriguing problem in genetics.

If our development was determined only by a genome with four different components, we'd all be about the same. The epigenome is the instructions on how to control the genome. The epigenome is responsible for activating and deactivating certain genes and programming the rate at which cells age. If every cell read all its genes and synthesised all possible proteins at the same time, the organism would not be able to function. Cells inherit not only their genome but also their epigenome. Figuratively, the epigenome is said to create the grammar that structures the text of life.

Neurophysiologist and epigeneticist Peter Spork says: "Change your lifestyle and you will start a chain of biochemical changes that will subtly but steadily help you and perhaps all your descendants for the rest of their lives on Earth. Although this statement sounds like what all the world's religions promise, it has a strict biological basis.

Genes can be likened to computer "hardware" and epigenetics to the software of the organism. Epigenetics is the science that studies changes in gene activity that do not affect structure. Epigenetics can explain how the environment can affect the switching on and off of our genes. Genetically identical laboratory mice were given different food during pregnancy. Mice born to mothers eating food with bio-additives were healthy and brown, while mice deprived of such food were born yellow and sickly. These changes further affected the animals' lives: poor nutrition switched off some genes in them that determine coat colour and disease resistance. The genes in the embryos at the time of feeding were already formed and were not affected - therefore something else was affected.

How do humans adapt to their environment in the long term? Previously, science knew only about two extremes - evolution, which takes thousands of years, and hormonal changes, which work very quickly. However, a middle mechanism has emerged in between - epigenetic switches. These are the ones that shape our adaptation to the environment and operate long term, even if no new signals enter the cell. This is why maternal nutrition or early childhood experiences can influence the rest of our lives.

But one should not think that the epigenome is an absolutely immobile system. A person is capable of changing the properties of his/her organism, both for the better and for the worse.

The influence of the environment and the epigenome has been better investigated in bees and ants. Bees initially develop as identical larvae. At the time they emerge from the eggs, nature has not yet decided which of them will be the mate and which will be the worker bee. Three days after hatching, the larvae of future queens are fed with royal jelly by nurse bees, while the larvae of worker bees are fed with common pollen and nectar. Feeding alters the epigenome. This was proven in 2008 by Australian researchers: they managed to get bee mothers without royal jelly, just by manipulating epigenetic switches. Now for the ants. The biggest of them all - soldiers - are three hundred times larger than the gardeners who tend mushrooms. Despite such differences, all these ants are brothers and sisters. The temperature and humidity of the place in which the ant larva develops is the decisive factor determining its future "caste". The ant's susceptible epigenome, reading environmental signals, switches on different genes, and the ant develops in one of the possible ways.

- Let me ask you one more question. Your explanations about genes do not quite agree with my ideas from the literature about genetic algorithms. If you could clarify, how do genes actually function?

- I understand your confusion. I'll try to tell you what's wrong.

You and I are building on a knowledge base of a physical-technical nature. Of course, there are nuances. You are more educated in maths and computer technology. I, on the other hand, knowing the basics of computer design, then read more about how scientists in physics and maths tried to understand the principles of life. Reflecting on this information led me to a curious inference: even the most gifted and well-trained physical scientists have a very, very simplistic view of life. For most of them, details are unimportant. They want principles. Newton, with his mathematical principles of natural philosophy, is a prime example of this. The trouble is that nature, and especially life, does not operate with principles. Diversity and confusion reign there. Moreover, two mechanisms, which, in fact, are antagonists, can coexist in one and the same organism until time. If the environmental conditions allow to survive and produce a viable offspring, it will pass on this illogical from the point of view of reason construction to the next generation. Such misunderstandings are most often encountered by physicians: in some conditions a drug helps, in others it may even have unfortunate consequences.

- I agree with you in general. But I don't see even a hint as to why the ideas of genetic algorithms specialists, which work so successfully in modern artificial intelligence technologies, do not reveal the true essence of genes' work. Please, as they say, closer to the body, Professor.

- Okay. Let me remind you that genes are in the nucleus of every cell. That's where they work continuously. It is this work that keeps the cell alive in the face of stochastic destructive influences. Remember this, a little later I will talk about how genes regulate the life of the cell. For now, let me explain that genetic algorithms have nothing to do with every second work of genes in each of our cells. The mathematical algorithms called genetic algorithms only reflect the authors' knowledge of the mechanism of evolution based on random mutations and inheritance of parental traits by offspring. This applies only to germ cells. The 218 other types of our cells have nothing to do with it.

So how does the cell genome regulate the state of the cell? This very interesting question is still largely unanswered. I will try to explain on my fingers what is more or less clear.

We have already said that cell specialisation is the result of the fact that in different cell types only unique fragments (sites) of DNA are actively functioning. There is such a concept as gene expression. It can be higher or lower. In the extreme case, when expression is zero, the site does not work. Experts already know that the concentration of certain chemical compounds in the cytoplasm can affect (activate or inhibit) the expression level of certain genes. This mechanism of chemical negative feedback makes it possible to regulate the influence on biochemical chains and the concentration of similar chemical agents (adaptation factors, regulators) in cytoplasm in the zone close to optimal values. It should be noted that everyday life of a cell in conditions of fluctuating concentrations of nutrients, products of vital activity makes it constantly interfere with the activity of certain genes. Malfunctions in this mechanism trigger a chain of transformations, which are often the internal causes of non-trivial diseases. Further I will explain how the inability of these mechanisms of cell self-regulation is compensated by additional mechanisms that integrate many organs and systems of our body.

CHAPTER 3

Conversation three

Multicellular functional systems

- Hopefully, you already have a general idea of how the diversity of cells that make us up is formed.

- I can't say that everything has become transparent, but it is much clearer than before the conversation. It is not quite clear to me why there is such a huge variety of cells. After all, a simple calculation of all possible combinations of 220 cells gives a number that is many times greater than the number of atoms in the universe! Unfathomable complexity!

After all, you and I know that at the heart of such a complex technical device as a computer are only three logical elements - "AND", "OR" and "NOT". Well, let's add to them the element on the basis of which the memory is built - we get four. Even if we add the physical elements that support input-output of information to this count, still, compared to our organism, the computer looks like a very simple object. Are we still so far from nature in our technologies?

- The fact that our technical achievements in terms of complexity are very far from the complexity of living objects, even from much more primitive organisms than ours, is a fact that should be accepted and realised. Perhaps your emotional perception of this fact is also caused by the fact that the words "neuron", "neural network", "intelligent device" appear very often in the technical literature of your professional sector. This gives mathematicians and engineers a false idea of real neurons and their networks.

I would like to point out that one of the famous couple Watson and Crick, who discovered the DNA double helix, Francis Crick, later became involved in brain problems and headed a research institute in Canada for a long time. I remind you, he was a physicist by training. And, as a physicist - Nobel laureate, he thought that the brain is just that organ, in which physiologists and psychologists can not understand, but the powerful mind of a physicist, finally, will break through the frequent myths. And what do you think? After half a century of studying the brain, he comparatively recently admitted that brain science has not advanced one iota. The brain is still a mystery.

You're right about the complexity of the human body as a system of 220 elements. But, as criminologists say, there are "mitigating circumstances"....

Some of them you should have guessed from my talk on ontogenesis. If you remember, I said that each of our adult cells originates from one of the three germ sheets. This suggests that the cell has a number of constraints on forming connections with the descendant cells of another germ sheet. These are the first.

Second. The relationship within any pair of cells can belong to one of four types:

1. The output product of cell #1 is used by cell #2 as a nutrient to carry out its metabolism. This type of relationship can be called the "producer-consumer" relationship. It is clear that in this case the metabolic rate of cell No. 2 will correlate with the metabolic rate of cell No. 1;

7. The output product of cell No. 1 is not a nutrient for cell No. 2, but stimulates one of its output functions (for example, if cell No. 2 is a muscle cell, the stimulant increases the contractile force developed by cell No. 2);

3. Everything is the same as in point 2 with the only difference that instead of stimulation there is suppression (inhibition*);*

4. Cell #2 is indifferent to the state of cell #1.

It follows from these two "extenuating circumstances" that the actual number of intercellular functional connections is much smaller than the terribly large estimated number that follows from combinatorics. But it is too early to rejoice. There are also "aggravating circumstances" in intercellular relations. Let us try to understand what this is all about.

Let me remind you that a cell is a complex biochemical machine, i.e. everything in it is deterministic. At least, this analogy is useful for understanding some functions of the cell and prevents drowning in a mass of specific biophysical, biochemical and physiological terms. Although biochemical reactions are a stochastic process, there are complex mechanisms functioning reliably in this machine. They regulate the rates of specific biochemical reactions, the physicochemical composition of the cytoplasm, the organised transport of the different ingredients of biochemical reactions to the reactor, the quality control of newly synthesised macromolecules, their elimination when defective ones are detected, and also carry out many ways of responding to changes in the pericellular fluid. It is already known that the number of proteins synthesised in the cell has long passed the hundred thousand mark and new proteins are being identified every day. Besides proteins, a huge number of other biologically active macromolecules are synthesised in the cell . And each of them is synthesised strictly in time in different phases of the cell cycle.

There are many specialised proteins on the surface of the cell membrane that act as chemical receptors for specific biologically active substances. These receptor proteins are interacted with by specific proteins within the cell. This interaction, which is still largely unclear, is capable of modulating each of the

first three types of cell-cell interactions highlighted above.
I may surprise you somewhat with the statement that, in my opinion, the cell as a chemical machine operating a huge number of molecules is even more complex than our organism, considered as a system of 220 types of cells. This is indirectly indicated by time. In total, it took about 3.4 billion years of evolution to create a cell equipped with a nucleus (eukaryotes). Of these, it took about one billion years to create and fine-tune the nucleusless cell (prokaryote), and the rest of the time eukaryotes were created on its basis. Compare 3.4 billion years with the 800 million years it took to create the great diversity of multicellular plant and animal organisms, including you and me, from eukaryotes. That's almost four times less time.
- The numbers are certainly impressive. But how do we know all this and how reliable is this knowledge?
- As for the age of our solar system and the Earth, fairly reliable data suggests that the Sun is about 4.6 billion years old, a third generation star, and the Earth is estimated to be about 4.5 billion years old.
Gradually, the hot Earth began to cool, solid crust appeared on the surface, water vapour condensed and water bodies and rivers appeared. According to the latest data, primitive unicellular organisms have already been noted in fossils formed around 4.2 billion years B.C.E.. As for multicellulars, the range of estimates for the time of their appearance is about 200 million years relative to the average date of 800 million years before our time. At least many fossils of invertebrate multicellular organisms have been found living in the bottom mud of the world's oceans about 600 million years before our time.
- When did land animals come into existence?
- About 400 million years before our time.
- That is, the entire evolution of animals does not exceed 400 million years. Does that include the dinosaurs?
- Why don't you include sea creatures in the list of animals? Because by the time they took their first steps on land, many complex aquatic species had evolved. In Morocco and southern England, a wide variety of trilobite species are found in the fossil record. This was all known before Charles Darwin and played a role in his theory of species evolution.
As for the "national animals of the Americas" - dinosaurs - their beginning is attributed to about 230 million years before us. And I think you know about their end, don't you?
- Yes. 65 million years ago. A ten-kilometre asteroid hit the Yucatan Peninsula in southeastern present-day Mexico.

- You're right about the date and place. But to be more precise, not only crocodiles and our rodent ancestors survived, but also some of the dinosaurs, which by then looked like birds.
- That's funny! Our ancestors used to be afraid of being eaten by a dinosaur, and today we eat the descendants of dinosaurs. So much for the vicissitudes of evolution. In a couple or three hundred million years, the descendants of today's despicable bugs will dominate our mutant descendants.
- Hollywood films haven't beaten you! You can expect a lot more from evolution. By the way, many people don't know anything about the zigzags of evolution. It's a common misconception that evolution is progressive and perfects the living world. Far from it. Evolution uses the diverse material provided by mutations to form another zigzag. The reverse, regressive course is not excluded.
- Okay, I'm not that dark. I remember that from high school biology.
But I'd like to hear your opinion that life is a way of preserving genes over time. Do you agree with that?
- In my opinion, this sounds witty, but misrepresents the point. You can't take some part of an organism and apply evolution to it. Life as we know it is based on the property of DNA that this double helix of RNA molecules knows how to reproduce itself. Theoretically, it is possible to imagine another version of life, where all the structures of the organism are constantly renewed and the organism is immortal. It is just that life on our planet was formed on the basis of this special molecule - DNA. Although experts are sure that about one billion years before the DNA molecule appeared, the transfer of information to offspring was based on RNA.
In this connection, it is important to say that this early method of reproducing genetic information made many mistakes, which contributed to the diversity of archaic life. And DNA duplication is not such a reliable way to preserve genes. But there's no reason to grieve about it. Without this flaw, there would be neither us nor the diversity of life.
- That was interesting. Let's move on. So far, the brain seems to be picking up and the conversation is shaping up to be interesting. Let's try to figure out if all 220 cell types have played an equally important role in our progressive evolution, or if there are cells that are particularly important.
- Come on!
I'll tell you right away - there are cellular specialisations that are momentous, and some that are so-so, ancillary.
Certainly, in the group of fateful cell types we should include those thanks to which our organism digests food. This should also include the group of cells

without which food could not be obtained. The cells and their communities that perform the function of recognising and evading threats are equally important. Finally, the cells that ensure the growth and reproduction of the organism are equally important. Let's try to understand how these cells and their systems work.

Although I don't think that an artificially modified human being needs such an inefficient and, let's face it, unworthy of an intelligent being digestive system, it is necessary to know it. At least, if your customers are seriously thinking about ridding humans of disease. Too many ailments are directly or indirectly caused by digestive problems.

Perhaps the main feature of the digestive system is that it is an anatomically complete system. This type of system is contrasted with functional systems whose organs may be located in different parts of the body.

Let us first consider the anatomy of the digestive system. Its schematic illustration can be found in textbooks (Fig.H). Specialists identify more than 30 specialised structural parts of this system. Their joint and consecutive functioning provides crushing of large pieces of food, its preliminary preparation for decomposition into chemical components, their absorption into the blood, filtration of useful components, as well as excretion of undigested parts of the food mass into the environment. By the example of the digestive system we can see that it is the result of a long evolution. This is better seen when you compare our digestive system with those of different animals of the tree of life. The general plan of the structure of this system is such that there is an internal environment - a cavity around which all consumers of nutrients - cells - are located.

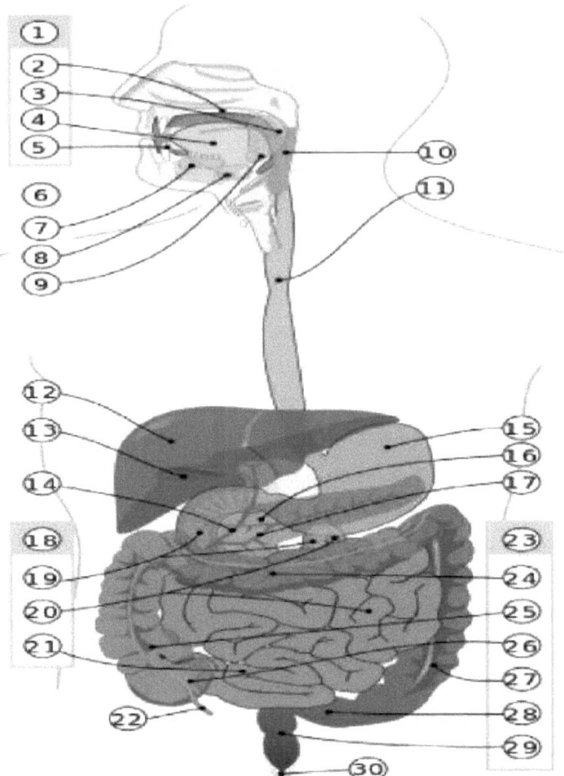

Figure 3. Digestive system

1. Oral cavity; 2. Palate; 3. Palatine uvula; 4. Tongue; 5. Teeth; 6. Salivary glands; 7. Sublingual gland; 8. Submandibular gland; 9. Parotid gland; 10. Pharynx; 11. Oesophagus, 12. Liver; 13. Gallbladder; 14. Common bile duct; 15. Stomach; 16. Pancreas; 17. Pancreatic duct; 18. Small intestine; 19. Duodenum; 20. The jejunum; 21. The small intestine; 22. The appendix; 23. The colon; 24. Transverse colon; 25. Ascending colon; 26. Blind colon; 27. Descending colon; 28. Sigmoid colon; 29. rectum; 30. Anal opening.

If we look at the picture in more detail, we can group the oral part (oral cavity, palate, palatine tongue, tongue, teeth, salivary glands, hyoid gland, submandibular gland, parotid gland, pharynx - 10 organs); the intermediate part (oesophagus, stomach, pancreas with two ducts - 5 organs); intestinal part (small intestine, duodenum, jejunum, ileum or small intestine, appendix, colon, transverse colon, ascending colon, blind colon, descending colon, sigmoid colon, rectum, anal opening - 13 organs). Functionally, the digestive system also includes the liver, gallbladder and common bile duct.

I have deliberately dwelt in such detail on the representation of the digestive system. It reflects two very significant features of our long evolution.

The digestive system is the most diverse in terms of the number of components (organs). There are different specialised cells in each of the identified parts of the system. The central nervous system (CNS) is also directly related to this system. It innervates all the secreted organs. In the brain there are many nerve clusters that control the flow of impulses originating in different parts of the digestive system. Every person knows how excitement, fear, and other emotions are felt by the "stomach and intestines." I won't venture to give you the exact number of cell types, but about 1/3 of the 220 specialisations are directly or indirectly related to the digestive system.

- I seem to understand your thought process. You mean that if we subtract these "digestive" cells from 220 types of cells, then the real number of heterogeneous cells, and with them the complexity of the system that ensures the functioning of the organism, is greatly simplified.

- Well, you're right. That's what I was getting at.

- But such neglect of the digestive system will be justified only if the future mathematical model integrating physiology and intelligence adopts a different concept of material-energetic support of cell viability.

- Agreed. But we must highlight another functional system that has historically arisen and evolutionarily selected for the organism's defence against the harmful influences of alien genomes.

- I get it. You mean the immune system. I used to be interested in that system. I even looked at a few publications on its mathematical modelling. But nowadays, all I remember is phagocytosis, lymphocytes and T-helpers....

- I won't go into the details of the organisation and functioning of our immune system, but some basic insights and features will be discussed.

The main feature of the immune system is that it is not anatomical, it is functional. Its elements are scattered in different organs of the body. For example, the lymphocytes you mentioned are a type of non-grained leucocytes, i.e. they are blood cells.

Human blood contains two types of white cells: T and B cells. The former are formed by the thyroid gland. The latter are produced in the bone marrow, which is the main organ of immunity. Along with the bone marrow, the second major organ of the immune system is the thymus gland (thymus). It consists of immature and undifferentiated cells - stem cells - that come to it from the bone marrow. In the thymus, cells mature, differentiate and eventually form T-lymphocytes, which are responsible for cellular immunity reactions.

The immune system reacts to foreign elements. Their list is very large. For example, viruses, fungi, bacteria, plant pollen, house dust, chemicals, transplanted tissues and organs, own mutated cells.
Each such object is a carrier of a foreign gene for the immune system, so their common name is "antigens".
Lymphocytes are also produced in the tonsils. They are located on the back wall of the nasopharynx in the upper part of the nasopharynx and consist of diffuse lymphoid tissue. One of the central organs of the immune system is the spleen. It receives arterial blood via the splenic artery to cleanse the blood of foreign elements and remove old and dead cells.
In addition to the above mentioned (central) organs of the immune system, there is also a peripheral system. It is represented in human organs and tissues by a branched system of lymphatic capillaries, vessels and ducts. The peripheral lymphatic system includes specific formations - lymph nodes. Their major part is located in the inguinal region, in the area of the axilla, at the base of the mesentery of the small intestine and others.
Lymph nodes are the "filters" in which pathogenic bacteria are destroyed. Lymph nodes are the guardians of lymphocytes and phagocytes. They form the immune response. The lymphatic system works in close relationship with the circulatory system and is constantly in contact with tissue fluid, through which nutrients are delivered to the cells. Transparent and colourless lymph transports metabolic products into the blood via the lymphatic system and is the carrier of lymphocytes, which come into direct contact with antigens.
It is believed that the structure of the immune system is only slightly inferior to the human nervous system in its complexity.
- All of this is certainly useful to know for understanding health and disease. But it seems to me that the problem of the immune system for the proposed mathematical model of an artificial organism needs to be thought through more than once.
- I am not the best adviser in this matter. If it comes down to formulating detailed technical specifications, perhaps I can do something to help. In the meantime, I would like to return once again to the previous question - what part of the 220 types of cells is involved in the immune system. I don't want to make a mistake, so I won't name the exact number. But it's a big number.
- It seems that few types of specialised cells are actually the main ones, and in addition to them evolution has "piled up" a much larger number of auxiliary cells, which are needed only in poor habitats.
- I completely agree with you. Moreover, this excess is even more noticeable when you compare the number of working and so-called silent genes. When I

started to take an interest in these questions from the point of view of energy efficiency of our organism, I came to the conclusion that, economically speaking, there are too many overheads. I remember once I even had an idea to propose to geneticists to remove everything unnecessary from the genome of some animal and see how viable such an organism is in ideal conditions, when the environment is sterile and the food is as diverse and nutritious as possible.

- So? Is anyone interested in the idea?
- Yaeyene has published.
- And I was interested in this! Indeed, what would happen to such a organism? Can we fantasise a little about it.
- I like to fantasise, for one thing. Let's give it a shot.

In my opinion, such an organism would function until the Hayflick limit is reached in the cells of critical organs (e.g. heart, kidneys).

- I think I've heard that name, but I don't remember exactly. Could you explain the gist of it?
- The biologist Heiflick, observing colonies of various cells under artificial conditions providing sufficient nutrition and removal of metabolic wastes, found that after a certain number of divisions the cells stop dividing and die. This number varies slightly for different types of cells, but on average it is about fifty cycles of division. It was later discovered that this pattern is caused by the fact that during each act of division, the chromosomes shorten slightly. More precisely, it is not the DNA-containing part of the chromosomes, but special end structures - telomeres.
- Oh, I remember now. It seems that an American woman was awarded the Nobel Prize for telomeres a few years ago. This theory was at one time widely discussed in the press in connection with a possible increase in life expectancy. I even remembered the word "telomerase". It's sort of an enzyme that restores telomere length. Isn't it?
- That's right! To be precise, the 2009 Nobel Prize in Physiology or Medicine was awarded to Elizabeth Blackburn, Carol Greider and Jack Szostak "for discovering how telomeres and the enzyme telomerase protect chromosomes."

Although many years have passed, there is still no correct medical technology to fight aging. Various types of medical entrepreneurs are siphoning off money from the rich who want to prolong their comfortable existence, but a growing number of scientific publications indicate that the case is probably more complicated than the telomere concept of aging and the Hayflick limit.

- What else did you want to say?

- In my opinion, such a hypothetical organism would be vulnerable for another reason. In order to understand it, we would have to delve into the principles of how each of our cells works.

- What do you know about the liver? It protects us too, doesn't it?

- You're right, of course. The liver protects us. But mainly from poisonous substances that are absorbed into the bloodstream during digestion. The liver cells - hepatocytes - make a complex chemical cocktail from these poisons - bile. This bile accumulates through the bile ducts in the gallbladder, from where it enters the intestine.

In part, bile aids digestion and useless ingredients are excreted with the faeces, giving it its characteristic colour. I mention this point only to show how doctors use information about stool colour to diagnose liver disease.

- And can you tell me what information a doctor gets by feeling the liver?

- First of all, this is how the doctor estimates the size of the liver. They roughly know the outer contours of the liver in the abdomen. Going beyond these limits signals that the liver is chronically overloaded and does not cope with its basic - barrier - function. That's what happens with poisoning.

But in regards to this question of yours, I would like to expand on the topic. The fact is that the actual size of any of our organs is made up of the size and number of its cells, as well as the vascular network of the organ. The enlargement of an organ means that it has either increased the size of each cell or created new ones. Of course, there can be a combination of both. But the important question is what size of an organ should be considered normal and under what conditions does hypertrophy of an organ occur?

- Are you saying there are some common patterns for all organs?

- I think I do. Let me try to explain.

After all, none of our organs is an independent object. Some organs are united in one functional system, others - in another. Conventionally speaking, for each organ there is its functional partner. In such a paired system, one organ is a loader, the other is a loadable organ. Say, for example, the liver is a loading organ and the gastrointestinal tract is a loading organ. In a state of health, the loading and the loaded organs must be in statics. In other words, the loading organ should have such a number of cells, which at their current size provides utilisation of products of the loading organ. If not all the production can be utilised, the residue activates mechanisms that enhance biosynthetic processes in the loaded organ. As a result, its cells increase in size, and increased proliferation increases their number. An external observer records these changes as hypertrophy of the organ.

- I wonder if this mechanism controls the physiological process of cardiac hypertrophy in athletes.
- Yes, that is true, although each body has its own nuances.
- Would you also say that senile or age-related prostatic hypertrophy is also of that nature?
- In general, yes. After all, the size of this gland is determined by the body's need for its production. Unfortunately, endocrinologists only know very roughly the ratios between the various hormones that establish the required size of this important organ. As the activity of some glands producing male sex hormones decreases with age, the normal ratio of all regulatory hormones is disturbed. In my opinion, prostate enlargement is an attempt to restore the disturbed balance of hormones.
- So hypertrophy of this gland is not always a disease?
- I think that this is one of the manifestations of an adaptive compensatory reaction to the reduction of secretion, nothing more. Unless, of course, we are talking about carcinogenesis. Even in cases of oncology in men aged sixty or more, oncologists themselves often say that a man would rather die with a diseased prostate than from it.
You know, our conversation today has gone on long enough. Let's continue it tomorrow. And to keep you thinking about this topic, as is the practice in multi-part films, I will end with something intriguing.
- If you're gonna do it, do it.
- Not everything in the body is directed towards the well-being of the cells! There is indeed synergy between individual groups of cells, but the relationship between other types of cells can't be called antagonism.
- Vague, but really intriguing. After all, I used to think that, like Dumas's Musketeers, the principle of "each for all, all for one" should work in the body. And you're saying that the cells are fighting each other. I hope you'll explain.
- Tomorrow, with a fresh head, we will look at the basics of biodynamics, the role of energy in it, and the general principles of any of our specialised cells. And then, I promise, I will clarify how cell synergy and antagonism coexist in one cellular community and how these cellular rules set the physiological rules of the whole organism.
- I agree.

CHAPTER 4

Talk four

Biodynamics and energy

- Are you ready to talk?
- All in all, I'm ready. But I would like to report that this morning I sat down and listened to all of our conversations recorded so far. I must say that certain moments of insight are already there, although I have not yet grasped the place and role of certain nuances.
- In my opinion, this is all fine. I have not yet fully talked about my vision of the organism, nor about important fragments of knowledge, without which your conversion to my faith is unrealisable. But let's take it one step at a time. We agreed yesterday to continue the conversation we started about the potential problems of a perfectly designed organism.
- I'm ready to listen.
- So, we are considering a hypothetical organism with nothing superfluous in its structure. I draw your attention to the fact that this statement also applies to each cell of the body: its genome must contain only working genes.

Evolutionarily, each of our cells is forced to exist in an unstable environment. As a result of long evolution, the cell has selected mechanisms that react sensitively to environmental changes and correct its metabolism in such a way as to minimise inevitable destructions. I would like to emphasise the word *"destruction"*.

Erwin Schrödinger, who is not unknown to you, published a book "Life from a physicist's point of view" in 1946. Generally speaking, many famous mathematicians and physicists of the mid-twentieth century tried to contribute to the understanding of life. Last time we mentioned that one of the fathers of computer theory, namely Alan Turing, also turned to biology and tried to mathematically justify the mechanism of formation of various patterns on the body of animals. But let us return to Schrödinger.

I recall his famous perplexing question: "I don't understand why one good biological molecule should be replaced by another of the same kind?".

He asked himself this question when trying to estimate the energy costs of biosynthesis. They turned out to be too high and in no way derived from the laws of physics and chemistry. For a long time this question remained unanswered. Only with time an acceptable explanation appeared. Its essence is as follows: biological macromolecules, most of which are tertiary or quaternary structures (there is such a classification in chemistry), are very unstable. They are sensitive to various destructive influences, including thermal vibrations of atoms. Since both reactions - synthesis and

decomposition - take place almost simultaneously in the cell, a much larger number of molecules have to be synthesised than would be required if each molecule were stable and useful for the cell. In fact, all biological structures and functions exist only because of a slight preponderance of the rate of synthesis over the rate of decay of macromolecules. This may seem paradoxical, but on the scale of organs and organisms, all this translates into the existence of many redundancies.
- I find this very interesting. And many people, especially engineers, are convinced that almost everything in biology is perfect and can be copied in technical systems. But here it is the other way round, isn't it?
I will later give some other examples to prove what I have said. You must understand and remember: the problem of survival is what has been and remains the main thing in biology. Perfection, optimality are not the criteria by which species were selected in the process of biological evolution. If there is a solution that allows to survive and pass its genes to the next generation, then the organism passes through the sieve of evolution. If not, such organisms are the end of the species.
- Brutal, harsh and instructive at the same time!
- Listen, I am pleased with your active and emotional participation in the conversation, but if you continue to react and interrupt me so often, I am afraid I will lose the thread of the narrative. And that will make it harder for you to understand.
- My bad, Professor! I'll try to refrain!
- I forgive you, let's move on.
If the rate of metabolism and with it the rate of energy expenditure (ATP molecules) increases, the cell must adequately increase the rate of ATP synthesis. This means that the cell must increase the total capacity of its mitochondria. There are several ways to do this. Firstly, special regulatory enzymes accelerate the process of ATP synthesis. Second, especially under chronic ATP deficiency, the cell increases the total area of its production units - mitochondria. There are two mechanisms for this too: hypertrophy of already existing mitochondria and
acceleration of their division (proliferation).
The relative contribution of the latter two mechanisms differs between cell types.
Similarly, if energy expenditures fall, the activity of mechanisms of struggle against energy deficit falls. More precisely, the above mechanisms cease to be active, and the natural process of macromolecule destruction does its job: over time, the new total area of mitochondria of the cell comes to such a

value, when the average rates of ATP synthesis and consumption are balanced.
We said earlier that activation of biosynthetic processes in the cell requires adequate activation of those organs and systems of the organism, which are somehow involved in increasing the total area of mitochondria.
Now let us consider another case. There is a balance of energy in a cell, with a high rate of expenditure balanced by a high rate of synthesis. What happens to this cell if the rate of energy expenditure falls to a minimum? As long as the rate of ATP synthesis is higher than the rate of consumption, the cell will not repair spontaneously degraded macromolecules in the mitochondria. Such a cell produces ATP only to supply its current metabolism. Over time, a new energy balance is established at a low level of ATP consumption. The energy capacity of the cell becomes low. Let's call such a cell a "squishy" cell. An organism made of squishy cells will itself be squishy. The first significant and sharp increase of load on such an organism will kill it!
In fact, I have here figuratively outlined the reasons why in each of our cells, and not only in our cells, most of the mechanisms are repeatedly duplicated.
You know that in technology, engineers duplicate critical components to improve the reliability of devices and systems. It turned out that in this case nature was far ahead of engineers. But any duplication is associated with an increase in overhead costs and a decrease in the efficiency of a given natural or artificial solution.
- I know that all duplication increases energy costs and reduces the efficiency of the system. Since nature has resorted to, or, to put it like you, put an inefficient organism through the sieve of evolution, there must be some explanation for this . Do you know one?
- Yes, I can explain that. But before I do, I'd like to make one important clarification.
When we talk about duplication in technology, we know that developers create two or more identical modules (unit, device) instead of one. All these modules can function simultaneously or can be switched on almost instantaneously when the main module fails. Here is a pure calculation of failure probabilities. It is well known that the probability of failure of two units simultaneously is much smaller than the probability of failure of a single unit.
This is not the case in natural duplication. There is no pure duplication in the organism. Instead, there is very often, almost at all levels of structures, more than one way of producing an output effect.
For example, you know that the energy raw material in the blood is glucose.

- ATP is derived from it.
- Yes, but not just from it. There are other ways: from other carbohydrates, from fatty acids, from fats. These sources of energy are accomplices of the normal process of ATP synthesis in cells, though they go through different biochemical transformations beforehand.
- I've even read that in cases of prolonged lack of food, the body's proteins are also used up. Is that where the energy comes from?
- Yes, it is. But I didn't talk about the norm for nothing.
Food is and always has been seasonal. If the organism extracted energy only from glucose, in the absence of mature fruits - suppliers of glucose, its chances of survival would be sharply reduced. But carbohydrates are found in many different plant foods. Therefore, those who were able to use other food resources for ATP production survived in evolution.
- Did they all have the same energy efficiency?
- You're in the top ten! No, of course not. Their efficiency varies, if only because the time it takes to assimilate them is different. But nature didn't shy away from the less efficient resources. She used them in the same way we use the elements to make batteries. The only difference is that, as a rule, in our artificial batteries, all cell sections are almost identical.
- Are you saying the battery principle is at work here?
- Yes and no. In nature, at least in the example of ATP synthesis considered by us , when summarising the effects of different mechanisms, we had to reckon with the asynchrony of some of them. In other words, in natural conditions there is an amplitude-temporal summation of the effects of different participants of a single process.
- I wonder if there are other examples to support such a conclusion?
- I may not know some things, of course, but I can give you a couple more examples.
For example, paired organs: kidneys, lungs, brain hemispheres. I apologise, the last example is not quite successful, as there are a number of significant structural and functional differences between both hemispheres.
- Can we include the paired sensory organs - ears, eyes?
- I think this example is not a good one either. In this case, the most probable reason for their presence is a new quality, stereo effect. The binocular vision provided acceptable accuracy in estimating the distance to a light source (direct or reflected), and hearing, through two ears, contributed to the accuracy of estimating the direction to a sound source. Given that in both cases it is about obtaining information about vital things (food or threat), the increased accuracy of estimates has played a role in the evolutionary

selection of organisms possessing such sensors.

- How does an ion pump work?

- It is a special enzyme (*ATP-ase) that* has the property of changing its spatial structure (*conformation*). The pump-enzyme works in three steps. In the first, it attaches three Na ions to the inside of the membrane$^+$. These ions change the conformation of the active centre of the enzyme, and hydrolysis of one ATP molecule occurs. In the second step, the energy released during hydrolysis is used to change the conformation of the enzyme pump again, so that three Na^+ ions and a phosphate ion leave the cytoplasm, Na^+ is detached, and two K ions are attached$^+$. Finally, the enzyme returns to its original conformation, the phosphate ion and K^+ ions end up on the inner side of the membrane. Here, the K^+ ions are cleaved off and the pump is ready to work again.

As a result, a high concentration of Na^+ ions is created in the extracellular environment and a high concentration of K^+ ions is created in the cytoplasm. This concentration difference is used in cells when generating or conducting a nerve impulse.

- What does the intensity of current ATP expenditure depend on?

- Intensity here should be understood as power. Consequently, it is necessary to take into account all kinds of biological work performed in the cell in a given time interval.

- What is biological work?

- It is a collective term that includes the creation and maintenance of a special chemical environment of the cytoplasm, the movement of molecules and atoms against gradients of their concentrations, the movement of ions through the ionic gates of the cell membrane against gradients of electrical potentials, and the restoration of the excitability of the cell after its solitary functioning.

- That is, we are talking exclusively about intracellular processes. But we also use energy for other functions, such as travelling in space. Why don't you talk about them?

- This is because any complex act, not necessarily a motor one, is made up of basic processes that take place inside a cell. Simply a complex act is always the result of functioning of chains formed from cells of different types. We will talk about it in more detail later. Here I would like to draw your attention to the fact that the energy expenditure capacities of different cell types differ considerably.

- I don't quite get it! Can you explain with examples?

- You've probably heard that the brain is a pretty voracious organ. I can give

you numbers. Weighing less than 2% of an adult's weight, the brain consumes about 20% of the body's total energy production. But there are even more voracious organs: kidneys, some organs of internal secretion. They consume even more energy per unit of mass than neurons.
- So what, let them spend it. What's wrong with that?
- Energy is one of the main resources that has been lacking throughout the evolution of organisms.
- Why, then, are we so proud of our wasteful brain?
- Desperation! The brain does not always work at full capacity. It, or rather, those of its structures that are involved in finding ways to solve non-standard tasks through associations with previous situations (tasks) and activation of logical thinking, tends, if possible, not to participate in the work of the body.
- I didn't expect to hear that from a physiologist! The brain organises and controls life activities, doesn't it?
- This is not the case. The brain is forced to activate when there is no automatic solution to a problem. Conscious search and development of a solution is the most costly part of the brain's work. Therefore, there are mechanisms that, after a solution has been found (for example, the development of skills), automate actions, transferring their realisation to the level of conditioned-reflex mechanisms.
- In other words - thinking is a luxury?
- Exactly!
- What then about other wasters, say kidney wasters?
- Unfortunately, the kidneys have to filter blood all the time. As they say, there's no bargaining here and we're not behind the price! But there is a mitigating circumstance: the mass of the kidneys is much less than the mass of the brain and the total energy consumption of the kidneys is not so great.
- And how can you indirectly judge which part of a person's body expends more energy and which part expends less?
- A good indirect indicator of this is the value of regional blood flow. After all, we have already mentioned that the raw material for ATP synthesis reaches the cells by means of blood flow.
- I want to ask another question: why is the body's energy consumption so high at rest?
- How do you know the body expends a lot of energy at rest? I didn't say anything about that.
- I read about it on some bodybuilding website. There, energy consumption was linked to the volumetric velocity of blood at the heart's output. Is that correct?

- Only with unchanged blood oxygen saturation.
- It has been said that the maximum heart rate reaches up to 200 beats per minute, whereas at rest the rate is about 70 beats per minute. Sports literature also provides data that at rest, the productivity of the male heart is about five litres per minute. At maximum load, the minute blood volume can increase 3-4 times. We also know that the maximum level of blood pressure in these conditions is only 80-90% higher than its level at rest. In general, there is a rather strange situation for the resting state: it seems that all cells are at rest (in technology the analogue is the idling state of the engine), but energy expenditures, even judging by the listed indicators of vital activity, are rather high.
- Good point! The standard explanation goes like this: these energy expenditures are necessary to maintain the basic metabolism of cells. We've already talked about metabolism. Indeed, biochemical transformations require a decent amount of energy expenditure. But it would be a mistake to write it all off.
Remember when I said that impulse generation in neurons and muscle cells involves ATP consumption?
- Yeah, I remember. But you said it was a small expense.
- It is really small if we restrict ourselves to the consideration of only one impulse. However, a single impulse arising in the soma of a neuron is further propagated along the axon. And it is sometimes quite long - more than one metre.
- What is the relationship between the path length of a nerve impulse and the amount of energy expended by a neuron? Is there a difference between the flow of current through an electrical wire and the flow of an impulse along a nerve? Does it take extra energy to propagate a single impulse?
- The mechanism of nerve impulse propagation is quite different from the propagation of potential and current in conductors.
The typical structure of a neuron is schematically represented in Fig. 4. Most large neurons have an axon, which is covered with a special insulator - myelin. This coating is regularly interrupted (approximately every millimetre) by a section devoid of myelin (Ranvier intercepts). In these intercepts, the impulse is re-generated. With an axon length of one metre, each pulse will have to be generated 1000 times. This is a considerable energy expenditure.
Now let us try to imagine tens of thousands of nerve fibres coming out of the neurons of the brain and spinal cord and going to the internal organs and skeletal muscles. Approximately the same number of nerve fibres carrying in

the form of sequences of impulses signals about vital characteristics of tissues and vessels of the body, goes in the opposite direction. This is already a huge number of impulses every second.

- Are there impulses circulating through all these fibre-channels even at rest?
- Exactly! True, the frequency of pulses at rest is several times less than the frequency at increased organ loads. , even at a basic frequency of 1-2 pulses per second, total

the cost of ATP molecules is impressive.

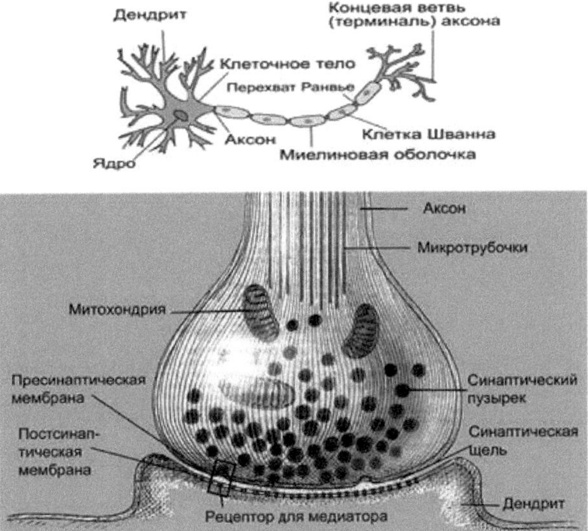

Figure 4. Schematic of the structure of a neuron (top) and synapse (bottom)

- I don't quite understand why neuronal impulsation is needed at rest?
- And then to ensure the functional integrity of all internal organs and the body.
- But this integrity is provided by cellular metabolites!
- Of course, the integration of cells by means of metabolites dissolved in body fluids, mainly blood, takes place. But this is not sufficient for an immediate response to sudden events, information about which is received by the brain through the sensory organs. It is necessary for the organs to be in basic (tonic) readiness. This is ensured by targeted pulses of low frequency to all organs. Without maintaining tonic activity of organ blood vessels, it is impossible to regulate blood pressure and cardiac output. Finally, to activate functional circuits (I will talk about them in detail later), the brain needs information about the current state of the executive organs.

Now I get it: the price to pay for the unity of trillions of specialised cells and

the responsiveness of a holistic multicellular organism is expensive.
- I'll tell you more: according to bioenergetics experts, a man of average build synthesises and spends about 40 kg of ATP molecules per day!
We will touch upon the issues of energy and biodynamics many more times. Then I will make the necessary additions and clarifications. I think this introductory information is enough to go further.
In my next talk, I'd like to talk about sensors.

CHAPTER 5

Talk five

Sensors

- If there are no questions on past conversations, let's continue.
- I apologise, but I didn't have time to listen to yesterday's recording again before coming in. I'll try to catch up tomorrow morning. If there are any questions, I'll ask them before the next meeting.
- I don't think it's critical to today's conversation.

So, as we agreed last time, today we're going to talk about sensors.

Everyone knows that we have five senses. It is through the organs of sight, hearing, smell, taste and touch (tactile sense) that our brain forms its internal model of the external world.

There are animals with a different set of sensory organs, hence their perception of the world is different. Here I want to emphasise that neither our senses nor the sensory organs of any other creature reflect all facets of reality. For example, with the help of physical instruments, we can track changes in electromagnetic vibrations outside the range within which these vibrations are perceived by the sensors - the rods and cones of our eyes. Similarly, a number of engineering developments allow us to indirectly convey information about the characteristics of infrasound and ultrasound, and the intensity of radiation.

I have listed just a few objective characteristics of the world around us. We have been immersed in this multifaceted physical reality since birth, and maybe even a little earlier. There is no reason to believe that the indicators of reality I have listed above - electromagnetic, radiation, sound - exhaustively describe the essence of the external world. It is just that we and other living organisms in the course of evolution have acquired such sense organs, thanks to which we managed to survive. In particular, the fact that our eyes are sensitive to the energy of electromagnetic vibrations in a very narrow range of frequencies (we understand them as colours) rather reflects the characteristics of the power of radiation from the sun. If it were a star with a different radiation spectrum, sensors of electromagnetic waves of a different frequency range would be required.

All that follows from this introductory sensory analysis is that the real world is likely to have some other view than the one perceived by us.

- I can imagine how this conclusion would be perceived by a painter. The vast majority of people in this profession, and of connoisseurs of painting, believe that what we see is reality.
- Let me repeat once again: for each of us reality is individual and is the

result of the activity not only of peripheral sensors (light-sensitive rods and cones of the retina, specialised receptors of other sensory organs), but also of the brain.

- If the perception of reality is individual, how do we communicate, exchange information about the objects of the external world?

- I cannot give an exhaustive answer to this philosophical question. I will only remind you how materialist philosophers said about it: reality is objective, given to us in our sensations and does not depend on us.

Perhaps, as a physiologist, I could ask a counter-question.

How do you think we perceive pain, emotions? I mean, you can't argue that they are individual. Nevertheless, people exchange about their emotional experiences. Moreover, when we go to the doctor, he is primarily interested in our subjective feelings, such as what hurts, how we feel. How can a doctor get information about the objective state of the patient's health from this subjective "porridge"?

- I don't know! But the doctor doesn't just use the history. He also measures something else, say, temperature, blood pressure, sends for tests. I think he was taught the general rules of making a diagnosis on the basis of instrumental data and the results of patient interviews.

- You're right about that - he's been trained. And this is diagnostic technology. I just wanted to emphasise that our daily activities are not based on precise information, but on some hints. And then it's tricky - our brain makes up and offers us its interpretation of the received set of fragmentary information of evaluative type. Strangely enough, this incompleteness of information does not prevent us from surviving and communicating.

- What about the brain's fantasies?

- It is up to each individual to put up with or fight this by-product of brain activity. If you are a writer and make up fairy tales with fantastic characters, your readers are unlikely to ask where you have seen them. They are gripped by the plot of your story.

Only one category of people - scientists, who, apparently, with imagination are not all right, are always trying to objectify reality. To the brain itself, all of its creativity seems real. It is an interesting information machine. Its food is information. The result of its activity is knowledge, or rather a version, a model of knowledge built on the basis of fragmentary input data. If the brain were a real tool for objectivising the world around us, it would hardly work in sleep.

- I think I agree with you. After all, our brains, trying to explain the incomprehensible phenomena of the world, came up with different theories. I

think that the first theory was magic: I am at the centre of everything and I can control everything at will. Only disappointment in magic led to the opposite theory - there are all-powerful gods, nothing happens without their participation, everything is their will, but they can be placated by sacrifices.
- To come to such inferences, you didn't have to study the principles of the brain. You just used one of its properties - logical thinking. As a physiologist, I try to draw on brain science as well. So here's another example.
What we see in our dreams is as real as our visual impressions. However paradoxical this may seem, it is a fact. Moreover, every time a person tries to remember past events, a sequence of two phases of a single process takes place in his brain. Firstly, the state of neurons of the visual cortex is recreated, which corresponded to the state of neurons during the primary act of vision. In the second phase of this two-phase process, the current state of neurons is re-recorded. Quite often, additional information can be wedged between the first and second phases. In other words, the memory often distorts the previous picture. You have probably caught yourself more than once thinking that you cannot be sure whether an event actually happened or whether it is a fabrication. In clinical practice, this brain plasticity is known to be used to erase painful memories and replace them with pleasant ones.
Before we consider how it turns out that our sensory organs allow the brain to build such a model of reality that helps us solve everyday problems and even experience aesthetic pleasure, I would like to tell you about other types of sensors that most non-biologists are not aware of. We are talking about a set of special organs that inform the brain about the physical and chemical state of the internal environment of the body.
It should be noted that unlike external sensors, whose operational centres are located in the neural networks of the cerebral cortex, the models of the internal environment of the organism are formed in the subcortical, evolutionarily older structures of the brain. That is why almost all work with these models is carried out subconsciously. Only in special cases (for example, in chronic organ pathology) some vague associative images from subconsciousness penetrate into consciousness.
The sensory organs of the internal environment are commonly referred to as receptor organs. The sensitive end of the nerve fibre of a specialised receptor neuron is located in the tissue. All internal receptors belong to one of two classes, chemoreceptors and mechanoreceptors. Each such receptor is a bipolar neuron, that is, it has a sensitive (peripheral) and a centripetal (going to the brain) part of the nerve fibres.
Chemoreceptors react to changes in the chemical composition of liquid media

in which the sensitive terminal part of the receptor neuron is located. Depending on their specialisation, there are chemoreceptors sensitive to the concentration of hydrogen ions, oxygen, carbon dioxide. Mechanoreceptors, on the other hand, react to mechanical tension in the sensitive apparatus of the nerve fibre of a mechanosensitive neuron. A distinction is made between stretch and pressure receptors. When the geometry of the sensitive apparatus changes, a graded (force-dependent) electrical potential arises in it. It is transmitted to the body of the neuron and, under certain conditions, causes the generation of a series of action potentials (impulses) that follow the centripetal branch of the neuron. As a rule, the frequency of centripetal impulses is related to the intensity of the primary stimulus irritating the receptor. All receptors have their own threshold and response frequency limits. Typically, between these extremes, the frequency of impulses is proportional to the amplitude of the stimulus.

Pain *(nociceptive)* receptors are a special type of receptor. They carry information to the brain about excessive stimuli, often associated with injury.

- Wait a minute. You say that each receptor has a different operating range. It's not clear if it covers the entire potential load range, or some part of it?

- You're a little ahead of me with your question. I just wanted to bring you to a very important fact. So far, when speaking about sensory organs, I have deliberately not said anything about the number of single nerve sensors that make up a particular specialised sensory organ.

- I'm not a biologist, but I know a little about the eye. You may remember that I've been working on pattern recognition. There are millions of cones and rods on the retina. Are you saying that all the rods and all the cones are the same?

- No. Moreover, the presence of intrinsic differences in a population of identical sensory cells is a fundamental prerequisite for adequate perception of continuous changes in the sensed signal.

- Explanation.

- I think you will agree with me that if we do not mean extremely small values of length and time (i.e., we do not consider quantum phenomena in the scales of Planck lengths and time), then processes proceed continuously. It is clear that having only one sensory neuron, it is hardly possible to convert the dynamics of a continuous quantity into a pulse code with the required accuracy. There are several reasons for this.

First, the frequency of nerve impulses in the bundle is limited by the biophysical processes underlying neuronal impulse (action potential) generation.

Second, the presence of an action potential generation threshold automatically cuts off all subthreshold signals.

Third, it is theoretically possible to build a sensor that covers the entire range of input signal variation. But this would lead to a decrease in the sensitivity of the sensor to the rate of change.

Evolution has left a compromise option. And it is this.

A real sensory organ consists of a large number of sensory cells working in parallel. But there is one very important natural peculiarity here. It stems from the nature of the formation of the population of sensory cells. They are all the result of successive cell divisions of previous generations. There's a time lapse between two mitoses. That's one. The cells of the population are dispersed in the space of the sensory organ. That's two. The spatial and temporal dynamics of random environmental factors (e.g., electrostatic field strength, radiation intensity) contribute to the formation of each cell of a new generation. These factors are non-stationary, so the indicated contribution for any pair of cells will be slightly different. There is one additional factor that causes cells from the same population to not turn out to be exact clones. This is the fact that the cell cycle has distinct phases, the duration of each of which depends significantly on the influx of nutrients with blood. Neither the blood velocity nor its chemical composition remains unchanged throughout the cell cycle. The result is that important biophysical characteristics of sensory cells such as response threshold and saturation level become individual for each cell.

- Are you saying that it increases the sensory sensitivity of the sensory organ to the dynamics of the input signal without increasing the sensitivity of each nerve cell?

- Exactly. You get the point. Another man would say, "That's how wise nature is. I'm saying no wisdom at all. The simple and effective solution is the existing mechanism of cell division. The zest is that the real process of multiple duplication of elements generates internal heterogeneities in them. The result is a population of sensors that react asynchronously to the dynamics of the environment. In other words, the dynamics of the input impact is encoded by a sequence of discrete sensor output signals. If at the receiving end there is such a device, which is able to differentiate the intervals between the incoming pulses, the task of the sensor system is solved.

- But to provide the necessary accuracy of discretisation of a continuous environmental parameter over its entire range of variation, many sensory cells are needed.

- That's the way it is - more often than not, millions.
- And how is the practical linearity of the organ's response ensured? After all, any specialist who knows the basics of measurement theory will tell you that the sensitivity of a sensory organ must remain constant within the operating range of changes in the measured characteristic.
- You're right again. In engineering this requirement is quite strict. But I am not aware of any technical measuring devices in which millions of sensors working in parallel would differ significantly from each other. In technology, there are clear tolerances in this respect: all sensors must differ from the reference sensor by a negligible amount. I want to emphasise the word *parallel* once again. *I was* talking about asynchronous triggering of different sensors in the population. There is nothing parallel here.
- Something I don't understand. What does this clarification have to do with the requirement of linearity of the sensory organ?
- Here we should say thanks to the great mathematician Gauss, who in his time established the normal law of distribution of natural random processes. You seem to have missed one important point of the process of forming an internally heterogeneous population of sensor elements. Wasn't it about the random character of the values of the factors determining the value of the response threshold or limit value for each newly formed sensory cell?
- Yeah, you've talked about those factors before.
- Well, doesn't it follow that in most cases (what medics call population norm or health) the statistical histogram of distributions of the number of sensory cells in an organ according to the magnitude of their response threshold will approach the bell-shaped theoretical curve, described by the Gauss formula $f(x) = \frac{1}{\sigma\sqrt{2\pi}} e^{-\frac{(x-\mu)^2}{2\sigma^2}}$? Recall that μ median (mean value), σ - standard deviation of the distribution, a σ^2 - dispersion.

Figure 5 shows two plots: the Gaussian curve at the top and the first derivative of this curve at different *σ im* .
- Well, such curves are familiar to anyone who has studied statistics.
- I'll say more. All researchers dealing with multicomponent objects encounter these curves. The abundance of these curves is reflected in publications on experimental physics, chemistry, biology, and even sociology and economics.

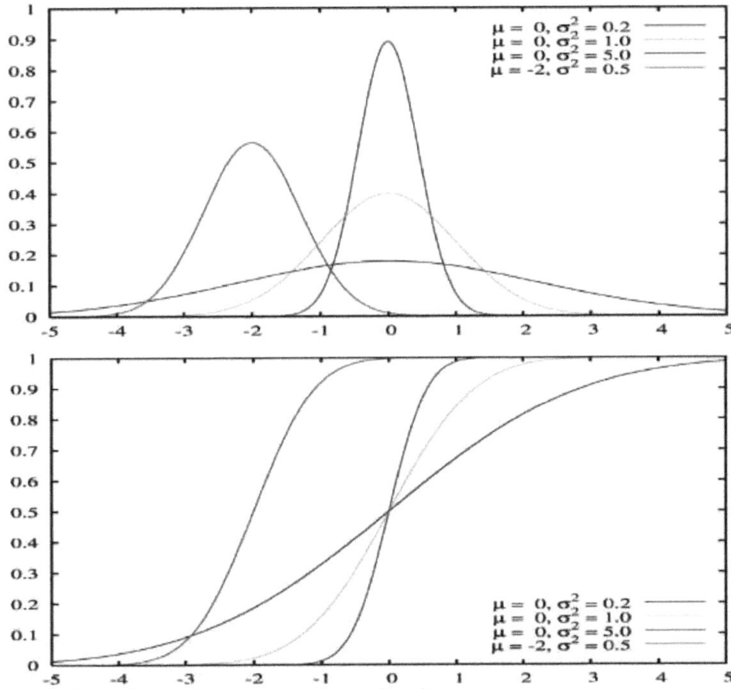

Figure 5. One-dimensional normal distribution

I think it is worth recalling that Gauss himself did not derive his famous normal distribution formula by theoretical means. He found a formula that approximated well the histogram of distributions of measurement results. The apple didn't just help Newton. Gauss had his own interest in apples. He did a special study: he weighed each apple of an apple tree crop. He noticed that the number of apples of different weights was different. He divided the entire weight range into equal intervals and counted the number of apples in each interval. From these measurements, he constructed a histogram and found that the highest number of apples were in the middle values of their weights. To the left and right of this weight value, the number of apples in each weight category decreased almost symmetrically. From this observation, he concluded that all natural random phenomena seem to have the highest probability of having mean values. And that's normal!

- That sounds quite funny. I have used the normal distribution formula many times and could not understand how the two famous numbers e and π appeared in this formula ! Now I see. He just found the right approximating formula.

- It's doubly funny because I've been thinking about this question too.

Let's put on the horizontal axes of the graphs in our figure the values of thresholds of sensory cells. And on the ordinate axis - the number of cells with this threshold. Let's take a look at the bottom figure. Firstly, as it can be seen from the graph, there are obvious non-linearities between the extreme values of the threshold, i.e. to the left and right of the middle. The middle part of the graph, on the other hand, is almost linear. In other words, the answer to your question about the constant sensitivity of the sensing organ is as follows: in the area of operating values of the organ, it provides almost constant sensitivity.

The second conclusion is that at extreme values of the input signal, the error of such a sensory organ increases due to nonlinearity. But I also want to note that the maximum sensitivity of the sensory organ falls on the average values of the threshold of the receptors.

Looking ahead, our sensory organ cannot be considered ideal. But, given that for most of our lives the magnitudes of external signals are also normally distributed, the organ and its associated response system are performing almost optimally.

- Look, you've completely convinced me. How long has this been known?

- Finally it's time to brag! The idea came to me at the end of 1988, but I proved it mathematically only in 1990. There was a corresponding publication in "Papers of the USSR Academy of Sciences". There were other publications. But I'm afraid that until now biologists are not even aware of the existence of this pattern. Their brains are structured differently. To make such an inference, they would have to test every receptor of all sensory organs! This is not only difficult, but it is not clear why.

- One more question, that's all. You didn't say anything about body hair.

- I didn't say anything about the scalp either. What of it? Do you want to know my opinion on whether humans will need hair in the future, or whether this vestige of the animal past will only provide sex appeal?

- Let me formulate the question. I consider body hair, at least on the extremities, as a certain resource, irrelevant at the present stage, which could serve as an additional channel for input of information into the brain. To clarify: this channel could be the basis for perception of such modalities of reality as, say, radiation and infrasound. Is this possible?

- Theoretically, yes. This is indicated by the fact that in some animals (e.g. cats, dogs, sea leopards, seals, and calans) a part of hair on the muzzle is thickened and serves as a vibration receptor. If terrestrial animals use these *vibrissae* to determine the dimensions of the openings through which they are going to crawl, the above-mentioned marine predators use them to sense the

wave trace left by fish. This organ turned out to be a rather effective evolutionary acquisition for them.
If this idea is properly developed, perhaps it could be useful for the human of the future. But I think it is more realistic to use such tactile receptors in robotics.
- Now rest, and I will briefly outline how I understood the structure and operation of the sensory organ. You will correct me if anything is wrong.
So, in some unknown way, a specialised sensory cell was born. It has a threshold characteristic - it is triggered only when the signal level at the input exceeds the threshold value. The input signal is analogue. The sensor produces impulses. In other words, we are dealing with discretisation of information about a continuous dynamic variable. The number of output pulses is proportional to the value of the input signal, but cannot exceed a certain limit value, which we call the saturation level of the sensor. A cell divides repeatedly according to its genetic programme. Over time, a whole population is formed from the original sensory cell. Due to slight differences in the conditions of birth and maturation of each new, daughter cell, the resulting population contains sensors that are not exactly identical. When a signal of variable amplitude is applied to the input of an organ, the number of responding cells varies with the speed and amplitude of the signal. All this allows the primary continuous dynamics of the signal to be encoded into a
as centripetal pulse sequences so that the number of pulses encodes the amplitude of the signal, and the interval between neighbouring pulses encodes the rate of change of the signal.
- Yeah, it's time to commend you, too. You got it right. Let's move on.
I will not dwell on the nuances of different types of our exteroreceptor sensory cells. I will only note that the method of encoding is based on the threshold characteristic of the sensory neuron.
Next, I think it is necessary to talk about interoreceptors. Let me remind you, that they supply the brain with information about internal parameters, i.e. vital physiological variables. There are many such variables. Since I am close to the problem of blood circulation regulation, I will outline the mechanism of functioning of mechanoreceptors located in the walls of arteries. With the help of these receptors, structures in the brain automatically adjust the value of blood pressure. Since these receptors are scattered in different parts of the arterial tree, it is appropriate to speak of a receptor system. The centripetal nerve fibres of different receptors are assembled in the form of a multifibre nerve.
There are specific nerve clusters (nuclei) in the brain, more precisely in its

part, the medulla oblongata, which controls the so-called autonomic functions. No one has exactly counted the number of nerve cells in an individual nucleus. But it is known that some nuclei mainly react to respiration, others - to blood pressure, others - to pain, others - to impulse streams coming from the digestive organs, as well as from special peripheral chemoreceptors reacting to the acidity of the environment, to the concentration of hydrogen ions or oxygen in the arterial blood.

There was a time when it was thought that one nucleus regulated the activity of the heart and blood vessels, another regulated the lungs, and so on. It is now known that this was a misconception. In fact, one ascending sensory nerve fibre forms many preterminal branches. They form synapses with neurons of several nuclei simultaneously. Moreover, part of the terminal synapses converging on one neuron may belong to the baroreceptor fibre, another part to the chemoreceptor fibre, and another part to the pain fibre. Moreover, nerve fibres capable of modulating (stimulating or inhibiting) the activity of these neurons of the same nucleus come from the neurons of the upper brain structures to the neurons of these nuclei. In general, it seems that each neuron of the considered nuclei is an integrator of information of different modality. In other words, the nervous regulation of organs and life-support systems is more complex than it was imagined a few years ago.

All this made me propose an alternative concept of blood circulation optimisation, which will be discussed separately. In the meantime, let's talk a little more about arterial baroreceptors.

The largest accumulations of mechanoreceptors are spatially localised in three regions of the arterial tree. One of these areas (zones) is located on the aortic arch and occupies approximately 1.5 square centimetres. The other two are located on the right and left *carotid* arteries at the point where the common *carotid* artery branches into external and internal branches (this location is called the *carotid sinuses*). The receptors respond to pressure differences between the inside and outside of the arteries.

Physiologists believe that the relative contribution of the receptor zone of the aortic arch is approximately equal to the total contribution of the zones of both carotid sinuses.

CHAPTER 6

Conversation six

Causal chains

- Today I want to talk about how life uses long, multi-step causal relationships.
- Aren't you saying that life, devoid of intelligence, knows how to plan multi-move combinations?
- That's one thing you'll never hear me say.
- What, then, should I understand by long causal links?
- I'm not sure if you watched American cartoons, or at least the Soviet series about Leopold the cat and the mice.
- When American cartoons appeared in our country, my daughter was already out of the age of watching cartoons.
- You should! You know, I was luckier in this case - I laughed heartily with my little ones at the adventures of cartoon characters.
- What if I didn't see it and laugh, I don't understand anything in your explanation?
- It will be a bit tight, but I'll try to explain at least with the example of Leopold the cat and the mice.

The mice fought the cat in a peculiar way. They arranged rather complicated constructions of links, each of which was well fitted to the previous and the next. In statics the construction looked absurd: there was no visible connection between the first and the last link. But every time the cat stepped on the rather innocuous first link, it worked like the domino principle, but only with a certain time delay for the completion of each intermediate act (for example, until the candle burns out). When the current act was completed, the next mechanism was triggered. The cat often looked at everything that was going on with interest, not suspecting how it would end. When the last act of this mouse scenario was completed, the cat would finally get a beating.
- I don't know what you're getting at.
- The point is that I didn't say the main thing: each next link in the long chain needed only a little nudge to trigger it. That little nudge is what it's all about.

Life is based on irreversible chemical processes. But inside the cell there are also quite a few reversible chemical reactions. Those reactions that lead to structural complication (against the second law of thermodynamics, i.e., towards entropy reduction) require energy. And as a rule, there is nowhere to get a lot of free energy. But often natural processes are accompanied by the release of a small portion of energy. It may be enough to cause some specific but small simple effect. Such reactions are common in inanimate nature. But in cells many one-act chemical reactions are united in such a long chain, when step by step the next small portion of energy released triggers and supports the next chemical reaction. The output of the chain is a specific chemical agent, which sometimes gives rise to another long chain of causation.

Note that the conversion of carbohydrates and oxygen into the end product, ATP macro-ergies in the mitochondria, is just such a causal chain, known as the Krebs cycle.

- How did such a complex chain come about?

- Judging by the fact that biochemists have described different versions of this cycle in organisms at different stages of evolution, one answer remains: by natural selection.

- What other examples can you give?

- Let's simplify the examples a little bit. Here's the leg. It is made up of the femur, shin bones, and foot. The femur and tibia bones have end pieces to articulate as hinge devices. Attached to each bone are tendons and muscles so that contraction of certain muscles causes changes in the mutual arrangement of the bones. It's quite visual. The upper limbs are organised on roughly the same principle. In other words, the musculoskeletal system is quite simply a biomechanical causal system.

But our bodies have other complexly arranged systems in addition to the biochemical multitudinous relationships.

- Which ones?

- Well, at least the locomotion system.

- What the hell is that?

- To put it simply, it's about orientation and movement in space. I don't think I need to convince you how important this system is in our lives.

The human body has no special advantages in this respect when compared with other animals. A child does not learn to stand on its feet and take its first steps until it is about a year old. Some animals do this a couple of hours after birth. Of course, standing and walking on two legs is more difficult than on four. But I wanted to focus your attention on something else, namely the fact that this process involves not only a large number of muscles, bones and tendons, but also a number of mediators. These include specialised receptor organs, nerve conducting systems in the spinal cord and special structures in the brain. Moreover, this system is functional only with the participation of interfacing systems that ensure the vital activity of the cells of all participants in a single integrative process.

Disruptions in the normal function of any of these links in the long chain give rise to specific locomotor dysfunctions. Some causes of locomotor dysfunction are obvious (e.g., muscle or tendon injury, vestibular tremor). Other disorders (e.g. tremor in cerebellar lesions) have characteristic symptoms and are easily recognised by specialists. But there are problems, the source of which is inadequate functioning of life support systems of locomotion organs. Such problems are rarely the object of study of orthopaedic doctors. In my opinion, this general shortcoming of modern medicine, divided into narrow specialities, does not contribute to a correct understanding of the physiology of long-term integrative processes.

- Why do you associate integration with time? I have always understood the word "integration" to mean any joint functioning of several organs.

- This is a simplistic view of integration. Equally simplistic is the notion that all integrative activity is organised and conducted through the brain.

There is a life support system. It's sometimes called the autonomic system. Be that as it may, without proper cellular life support, any functional integration very soon disintegrates.

- Why?

- Because the self-interest of the cells and the whole body rarely coincide!

- That sounds almost provocative! I always thought cells were the body's servants. And here you're insinuating that they can harm each other. What, the cells are the robbers and the brain is the terrorist?

- It sounds funny, but there's something to it! Let's talk about that another time. Now it is more important to understand the principle of formation of causal chains from different cell types.

- Maybe what I'm going to ask is not really relevant to the topic of today's talk, but I'm somehow confused about the similarities and differences of specialised cells. Can we talk about this briefly?

- I'll give it a try.

For this purpose, let us return again to the conventional cell, which was taken as the foremother of all our cells. It was able to create such a physicochemical composition of the cytoplasm, which provided the optimal rate of its metabolism. I note that this composition differed from the composition of the pericellular space. I will also note that spontaneous, not very frequent external influences, here and there opened pores (ionic gates) of the cell membrane. The difference in concentrations of substances on both sides of the membrane led to ion currents through the membrane and the resting potential of the cell decreased. The cell fought this phenomenon with the help of special pumps located on the membrane. The pumps, powered by

ATP energy, soon restored the optimal composition of the cytoplasm. In other words, this composition usually fluctuated.
The vast majority of our specialised cells continue to exist in this mode. The only way of intercellular communication is to enhance or inhibit the metabolism of the partner cell by means of its waste products (metabolites). We already know that the medium for such communication can be fluids (intercellular, blood). We also know that such interaction does not have a high speed.
Now let us imagine that a certain set of mutations in the genome of this cell leads to the appearance of a new variant of the cell. In it, under the influence of external perturbing factors, many more ion channels are opened than in the original variant of the cell. What happens in this case?
- The concentrations of the same ions on both sides of the membrane will very soon become the same.
- Exactly! But that's cell death. Its metabolism will stop working. Therefore, it is logical to believe that only those cells survived in which, along with the above mutations, there was also one that contributed to the processes of restoring ionic disequilibrium on both sides of the membrane.
- The hint was understood: in addition to the process of concentration equalisation, the restoration of the initial disequilibrium of these concentrations also appeared. Since we are talking about ion concentrations, this two-phase oscillation looks like an electrical impulse. In fact, a nerve cell was born.
- You're right, but partially. A group of cells with this property has emerged. As I said, this group includes all kinds of excitable cells.
- What are the differences between the two?
- There are many differences, but so as not to overwhelm you with unnecessary details, I'll give just a few examples.
The first is cardiac muscle cells. Their characteristic feature is the long duration of the state of excitation - up to several hundred milliseconds. For comparison, the typical duration of a nerve impulse does not exceed a couple of milliseconds.
The second is pacesetter cells. So far I have said that the resting potential in nerve and muscle cells remains constant until an external factor acts on the cell to open its ion channels. Pacemaker cells differ from other excitable cells in that no external force is necessary to open their ion channels.
- From my student years I remember a special technical circuit called "multivibrator". It automatically generates pulses with a frequency set by the circuit parameters. Does the pacemaker cell fulfil the same function?

- Yes. Each pacing cell spontaneously generates an action potential. It is this type of cell at the entrance to the right atrium that gives the heart its spontaneous rate of contraction.
- How, then, does the rate of contraction of the human heart change?
- The point is that the rhythm of the pacemaker cell discharge is modulated depending on the temperature and chemical composition of the local intercellular fluid. It is due to the sensitivity of pacemaker cells to chemical agents that the interval before the beginning of the next act of heart contraction increases or decreases. As for temperature, when the blood temperature increases by one degree, the frequency of heart contractions increases by about 10 beats per minute. This is the neurohumoral regulation of heart rate.
- Where else besides the heart do pacesetter cells exist?
- Predominantly in the brain.
- How does a muscle cell work? You said it's also a member of the excitable cell group.
- At first, it should be clarified that there are three types of muscle cells. The first is smooth muscle cells, the second is skeletal muscle cells, and the third is cardiac muscle *cells, cardiomyocytes.* The muscle cell is oblong. In the transverse direction of the cardiomyocyte and skeletal muscle cell, dark and light areas alternate, so these cells are called transverse striated cells.

A characteristic feature of a muscle cell is its ability to shorten its length. This happens thanks to two intramuscular filamentous proteins, actin and myosin. These proteins are arranged in a single straight line. Unlike a smooth muscle cell, in the relaxed state of a transverse striated cell the actin and myosin filaments are partially overlapped, which creates the dark band.

Another characteristic feature of the muscle cell, which makes it similar to a neuron and at the same time different from it, is that after the state of electrical excitation, a special process of electromechanical coupling develops in the muscle. In this phase, actin and myosin filaments slide towards each other, their total length decreases. This is the effect of muscle contraction. It lasts for some time, after which specific ionic processes (I will not go into details) restore the initial state of the actin-myosin complex, i.e. the muscle relaxes.

Blood vessels are equipped with smooth muscle cells. The contraction of these cells reduces the vessel lumen. This property of smooth muscle cells is used by nervous and humoral mechanisms to control blood flow through regional vessels. Indirectly, changes in the length of vascular smooth muscle alter blood pressure.

- What about secretory cells? What causal chains are they involved in?
- This class of specialised cells is the most extensive. Suffice it to say that the digestive system is dotted with different varieties of these cells. Some produce saliva (a liquid with many food enzymes), others produce gastric juice, while others, located in the pancreas, produce insulin, which helps convert excess glucose into forms that can be stored in reserve in the liver and muscles. Each section of the long gut has its own specialised secretory cells.
- And they all had their own evolution too?
- I have no information about the evolutionary series of digestive secretory cells. However, there is reason to believe that the answer to your question is affirmative.

In general, we should not forget that the cell secret itself is not needed by its producer. In this connection, we would like to remind once again that each cell removes unnecessary metabolites. Since some of these metabolites turned out to be useful for cells of another type, a producer-consumer relationship is formed between them.
- But if the cell itself gets rid of unnecessary metabolites, why is the arrival of a triggering impulse to the secretory cell necessary?
- Normally, inside the cell, some metabolites accumulate in the form of bubbles. When there are a lot of them, there is a natural sluggish release into the near-cellular environment. But an external shock favours the simultaneous release of a large number of metabolite molecules. In addition, the accumulation of most metabolites in the cytoplasm by chemical negative feedback inhibits their production. Therefore, removal of metabolites from the cytoplasm favours chemical reactions that produce these metabolites.
- You haven't said anything about the skin, which is such an important organ. It is not for nothing that it has the title of the largest organ!
- The evolutionary versions of skin are countless. The general role of any kind of skin is to protect against physical damage and penetration of foreign, harmful microorganisms into the organism possessing the skin. Even plants have skin, or rather, bark with the same functions. If we narrow down your question to humans, the skin has another important function that can be considered from a causal perspective. I am referring to our ability to regulate body temperature, or more precisely, to maintain a relative stability of internal temperature. In scientific terms, this is called homoeothermia, which is opposed to poikilothermia (commonly referred to as cold-bloodedness).

The causal chain underlying the ability to homothermia involves almost the entire organism. This assessment is legitimate, if only because the heat to be distributed to different parts of the body is produced by all cells due to the imperfection of the mechanisms utilising the energy of macroerg decay. Further, in favour of this estimation speaks the fact that the heat allocated in separate regions of the body by means of blood is spread all over the body, i.e. there is an equalisation of temperature in actively working cells and passive cells occupied only with their metabolism and reproduction.

The skin, due to its large surface area, serves as a radiator. Through it the excess heat is transferred to the environment by radiation and convection, if, of course, its temperature is lower than the skin temperature. Otherwise, heat removal is carried out only by evaporation. Moreover, evaporation also takes place through the lungs.

I have already said that the increase in blood temperature by purely biophysical mechanism accelerates the work of the heart. In other words, the volumetric velocity of blood and its ability to heat and mass transfer increases. There is also a biophysical mechanism of vascular dilation from temperature. Both of these mechanisms accelerate heat dissipation.

Along with these biophysical mechanisms, there is a feedback-based regulatory system. The central link in this regulatory chain is a neuronal nucleus in the hypothalamus (a formation in the central part of the brain). These neurons receive information from peripheral thermoreceptors via afferent pathways. If the incoming impulses signal an increased temperature, the nerve centre sends inhibitory impulses to sympathetic neurons of the medulla oblongata. A drop in the frequency of vasoconstrictor impulses travelling from the brain to the skin small vessels (arterioles) causes them to dilate. This increases skin blood flow and heat transfer to the environment.

- How does the body fight the cold?

- Worse!

The human body is practically unable to control the process of heat production. It only combats its excess more or less effectively. Only in extreme cases of a drop in body temperature (below 32 degrees Celsius) muscle shivering is activated, but the additional heat produced is small. It is therefore necessary either to avoid frost or to insulate the body well from low temperatures.

CHAPTER 7

Conversation seven

Causal chains (continued)

- Yesterday we talked about multi-stage biochemical and physiological causal chains. Before we continue this topic and talk about other examples of such chains, I would like, as has become a good tradition of our conversations, to ask you if there are any additional questions for me?

- Yes, there are. When I listened to our recordings again last night, I clearly had a question: are there any incomplete biochemical chains in our body? To clarify: by an incomplete chain, I mean a short or long chemical chain that produces a product that is useless or even harmful to the body.

- I haven't read about the existence of such chains. But, on the other hand, there is no prohibition on the existence of such chains.

I'll say more than that!

After all, the cause of the appearance of any new branch on the tree of evolution (here I am talking about biochemical evolution) is a random process. During the life of an organism, a cell divides many times. Each act of division contains DNA replication. We have already said that this process does not have absolute reliability. Let me remind you that the average frequency of mutations in the human genome is estimated as 1 mutation per 1000 genes. In other words, about 25 mutant genes per each division. If a cell lives up to its Hayflick limit, this number is multiplied by 50 and we get 1250 mutations.

- Wait! But that suggests that in the old organism, almost 5% of the genes are already mutants!

- Although this linear extrapolation is not always accurate, the estimate is generally correct.

- And how many genes must undergo mutation for cancer to occur?

- This is where you expressed everyone's natural fear of cancer!

In fact, the role of mutations in the development of oncology is not as clear-cut as it is in the public mind.

- What can you do to allay my fears?

- I note that with your question you are leading me away from the plan for today's conversation. But since you insist, there is something I can enlighten you with.

There are many varieties of cancer, and each can have its own both causes and dynamics. Perhaps the place to start is with the fact that a normal cell grows and multiplies until it comes into contact with neighbouring cells. As

soon as it senses contact, it stops growing in that direction. If there is no more space around the growing cell, it stops growing and dividing, entering the *interphase of* the cell cycle. This phase is where most of our specialised cells can stay until some factor reactivates the cell division process.
In the aspect of oncology, it is important to remember that cessation of growth and division by a cell when it comes into contact with its neighbours is a normal phenomenon and is called *contact inhibition.*
Any oncology develops only after the mechanism of contact inhibition is disturbed. It has been established that this requires specific mutations in two genes, P16 and P53, to occur in the cell. The course of the malignant process depends on many additional conditions of life support of the mutant cell.
- Is this genetic change accompanied by any biochemical changes that could be markers for early diagnosis?
- I am not an expert in this field to name specific markers. But I have read some literature, so I will say that there are some. Moreover, the healthy immune system of the body uses a number of such markers to recognise and destroy a malignant cell.
- Is the emergence of mutant cells a rare event?
- I wouldn't say that. According to various estimates, at least 25 per cent of newly-emerging cells are not without flaws. Of those, up to 10% are potential cancer cells. Since you and I have reached a respectable age without cancer, there is every reason to believe that our immune system is still coping with one of its main tasks - fighting cancer cells.
- Why, then, do doctors scare us so much about carcinogens?
- Actually, doctors are not evil people! It's just the way they were taught by their professors. In the days of their youth and professional development, there was a popular theory stating that cancer is a consequence of almost any genetic mutation. Since a great number of chemicals have the property of causing mutations, they were classified as potential carcinogens. This also includes physical energy effects (ultraviolet, solar and cosmic radiation).
I even read a publication once that claimed that any wound that doesn't heal for a long time increases the likelihood of cancer. There is some truth to this, as wound healing is the replacement of wounded cells with new ones. With chronicity, a large number of new cells need to be produced. Increased cell cycle frequency increases the number of mutant cells and the burden on the immune system. But there is one more thing to note here.
The mutational theory of carcinogenesis believes that a single mutation can so change the biochemical processes in the cell that the probability of losing the property of contact inhibition increases over time. Personally, I am more

sympathetic to another theory, which was proposed by the American oncologist Peter Duesberg. He found that almost all cancer cells have highly altered chromosomes, which he called chimeric chromosomes.
To make it clear to you how Duesberg's theory differs from other, mutational, theories, I will note that on average there are about 1000 genes in one pair of chromosomes. A mutation is a random process that affects a small section of DNA. As a rule, a single mutation hardly changes the number of genes in a chromosome. In actual cancer cells, chromosomal aberrations occur, resulting in an increase of 600800 genes in the length of the chromosome. In fact, chimeric cells appear in the human body as a result of chromosomal aberrations. Moreover, not all chimeric cells are necessarily malignant. If you are an attentive observer, you must have noticed that with age, pigmented formations, warts and other parts of the human body appear on the face and other parts of the body. These are chimeric cells. Their genome differs significantly from the human genome. Nevertheless, we get along with them without any problems.
In this connection, I find one phenomenon amusing and instructive at the same time. Biologists once classified vertebrate animals by common external features. And it turned out that according to some features, a simple aquatic organism called a lanceolate is the ancestor of almost all vertebrates, including us. And recently geneticists, having compared the genome of the lanceolate with the human genome, were amazed: it turned out that our genome is almost a quadrupled genome of the lanceolate.
- Do you mean to say that some factor at one time caused such a chromosomal aberration in the Lancetnik that the genes overlapped in such large chunks and produced an entirely new organism?
- Actually, I'm not at liberty to make such sweeping statements. But so say the geneticists themselves. I only agree with them that four hundred million years is clearly not long enough for single mutations of low frequency to transform primitive simple organisms into the animal species we know. There is no alternative but to accept that from time to time the measured course of evolution was suddenly altered, and new species of organisms leapt into existence. In my opinion, the mechanism of chromosomal aberrations provides us with a simple and logical explanation for both speciation and oncogenesis.
- You see what a generalisation you've made. And I didn't want to deviate from the intended topic! I think that this seemingly private knowledge is not superfluous for a correct understanding of the rules of coexistence of our cells. It confirms that chance played an important, if not the most important

role in the origin and evolution of not only the simplest unicellular, but also multicellular organisms.
- On that optimistic note, let's return to our causal links.
Accepting that incomplete biochemical circuits in our bodies are not excluded, we arrive at an important philosophical question: is our body optimal if we look at it as a biochemical machine?
Much depends on the answer to this question in the philosophy of medicine.
If an organism uses resources to support useless biochemical transformations, it is clearly suboptimal. This raises new questions. How should we understand health? How should physicians strategise to treat disease?
Theoretically, the identification of incomplete biochemical chains conceals new avenues of treatment based on technologies to neutralise or use endogenous (produced by our cells) chemical products.
- That sounds too fantastic. Is there a more down-to-earth application for such knowledge?
- I intuitively feel that there is. There are even some considerations.
- Share.
- No, I won't. I need to think this through a little bit more
Let's go back to the problem of cell antagonism. Remember, I hinted at it at the end of our third conversation and promised to talk about it in more detail later. Some of these details have already been revealed when we dealt with the principles of sensor operation and considered causal relationships. I think my task is a little easier now. I should just make some generalisations.
In fact, we already know that each cell, regardless of its specialisation, is only interested in living as comfortably as possible. For this comfort it competes with all the cells in the community, i.e. it tries to increase its consumption to the genetically determined maximum. If it succeeds, it will limit itself to fulfilling its genetic programme of reproduction.
This maximisation is hindered by two circumstances: the low concentration of all vital substances in the blood and the small value of local blood flow. When all necessary chemicals are present in the blood, their redistribution in favour of one or another group of active consumers is carried out by physiological mechanisms. They regulate cardiac output and arterial pressure, dilate or constrict regional vessels. The deficiency of a particular substance in the blood can be replenished in two ways: from internal depots (if any) and by activating two specific causal chains. The first of these, the search for food, is a complex of behavioural reactions. The second, activation of the digestive system, combines behaviour and physiology.
It is known that the average metabolic rate, hence the average nutrient

requirement, is different in cells of different specialisations. It is also known that each cell's need for nutrient supply is variable, depending on which phase of the life cycle the cell is in. Consequently, in order to approach the regime, which was called comfortable above, some physiological mechanisms must continuously receive information about the needs of cells and try to satisfy them.

We already know that any more or less complex physiological mechanism, at least, consists of three obligatory functional links. The first of them is receptors, the output activity of which reflects the intensity of a controlled physical, chemical or biological parameter. The second is the central analyser link, which produces a corrective effect on the state of the third link - the effector. A remarkable feature of all the links is that their constituent specialised cells are forced to react to the external influence in a specific way. This specificity determines what is usually called "the output function of the specialised cell". But we already know that this function contradicts the original interests of the cell and disturbs its comfortable mode.

So, in order to preserve anatomical integrity and functional integrality of the organism, there is no other way but to periodically disturb the cellular comfort of all participants of joint functioning. This is where cellular antagonism lies. It destroys what each cell diligently creates for itself. **Paradox: the unity of cells is destructive for each of them!**

But it is precisely because of this fundamental paradox that any multicellular organism can exist. It turns out that the basic causal chains based on antagonism of cells must be combined with their synergy aimed at comfortable life support of each cell throughout life.

- Do you mean to say that this co-ordinated existence is ensured by special functions of the brain?

- Not at all, if only because this principle exists in many organisms that do not have a brain. On the contrary, I just want to lead you to the idea that each of our cells is able to temporarily escape from the influence of those influences that make it fulfil its role in chain (ensemble) functions. Just thanks to its egoistic mechanisms, each constituent cell of such a chain manages to restore its cytoplasmic comfort.

- Figuratively speaking, in order to become a soldier of the motherland, one must not only grow up and mature, but also take care of oneself from time to time. Am I interpreting your thought correctly?

- Quite.

Unfortunately, this fundamental view of the body as a specific community of different types of cells is not yet available to medical practitioners. This is

why they often treat some organs while damaging others.
- There's something else I wanted to ask you. What do you mean by "regulation"?
- I'll start with a formal answer. Imagine a function of two arguments.
- Pictured.
- Also imagine that changing these arguments by the same amount causes the function to change differently.
- And it has presented. We are talking about the sensitivities of a function to changes in arguments.
- Good! Now imagine that one mechanism controls changes in the values of the first argument, and another one controls changes in the values of the second argument. Then we can say that the resulting value of the function is regulated by these mechanisms. But this regulation is in no way tied to the current values of the function, because the regulatory mechanisms do not receive information about the function. Let us go further and suppose that some meter is able to register the values of the function and transfer this information to those mechanisms that increase (decrease) the values of the arguments. Now we are dealing with a closed system in which each value of the function corresponds to a pair of values of the arguments. In a real system, this correspondence will take place only in a limited range of changes of the function, because each of the arguments is limited in its changes.
If information about changes in the value of a function is used by said mechanisms to prevent these changes, such a regulator is said to operate by negative feedback. There are many regulators of this type in the organism. Within each cell they stabilise (experts use the term "homeostatize") the physical and chemical composition of the cytoplasm. On the scale of the organism, homeostatic regulators maintain relative constancy of body temperature, blood pressure, physicochemical composition of intercellular fluid and blood. In fact, homeostatic regulators reduce the sensitivity of natural life-support systems to abrupt, often destructive changes in the environment.
- And what types of regulators are available?
- According to the theory of automatic regulation, there are proportional, differential and integral regulators. All others are obtained from combinations of these regulators.
- No, you misunderstand me. I know this basic cybernetic theory. I'd like to know what types of regulators there are in the body.
- That's what I would say! Nature uses only proportional regulators, i.e. the

control action is proportional to the amount of mismatch between the desired and real values of the regulated function.
To conclude the conversation about causal relations, I will say the most important thing: none of the participants of multicellular causal relations (in particular, cells) "wants to work for others". Nevertheless, evolution has preserved those multistage structures and systems, the result of the functioning of which was beneficial for the organism.
- To go to a philosophical generalisation, life is a by-product of useful combinations of random phenomena and mechanisms.
- Although it sounds paradoxical, that's life!

CHAPTER 8

Conversation eight

Problems of cellular life support

- I hope you already have an idea of the basic principles of how a community of cells functions?

- What I realised is that they are selfish, compete for food, can cause problems for other cells, and with some of them form functional systems that serve our interests.

- This is already significant progress. But we still need to understand why the joint evolution of cells of different specialisation did not contradict the goals of the basic mechanisms of cell metabolism. After all, without maintaining an optimal level of metabolism, cells become stunted, which makes the organism unhealthy.

- If you can't do without food and water, then tell me how the body manages to solve the most urgent problems of cellular life support?

- I will again start from afar and present the mechanisms in a simplified way.

Let us assume that our distant unicellular ancestor already knew how to solve its basic tasks quite well - assimilate nutrients, remove pollutants from the cytoplasm into the nearest pericellular space (liquid) and reproduce.

Now suppose that such a cell is surrounded on all sides by sister cells, with intercellular fluid between them. As long as the number of cells in the community is in the tens and hundreds, the diffusion rates of oxygen and carbon dioxide and nutrients allow the cells to maintain their leisurely metabolism. But even in this autonomous community of cells, some cells receive more of the necessary chemical ingredients than others. This basic heterogeneity (cells would say "social injustice") will not be eliminated even if evolution provides other mechanisms for nutrient (nutrient) supply and waste disposal.

- I get the impression you're not going to talk about the living cells of our body, but something else. I have a feeling it's going to be a plumbing problem!

- Your gut doesn't deceive you. The plumbing problem is not just a civilisational problem. A community of cells had to successfully solve this problem in order to survive. There are several known solutions. But we will consider the one used in our body.

Here we need to separate the problem of nutrient delivery from the problem of providing the proper chemical environment of the cytoplasm.

The network of arterial and capillary vessels serves as a transport system for delivering everything each cell needs. Arterial blood carries chemical compounds absorbed into the bloodstream from the intestine and oxygen to the cells. Some oxygen is found in the blood plasma, but its main carriers are red blood cells - erythrocytes.

The red blood cell has a protein called haemoglobin. Depending on the conditions, its ability to combine with oxygen (affinity) becomes greater or lesser. Oxygen as a constituent of air enters the pulmonary alveoli (peculiar air sacs with a thin membrane permeable to gases). Alveoli are enveloped by a network of pulmonary capillaries, in which the concentration of carbon dioxide is greater than in the alveoli. Therefore, carbon dioxide retained by haemoglobin is released and diffuses along concentration gradients into the alveoli, where its concentration is low. Conversely, the concentration of oxygen in the alveoli is higher than in the pulmonary capillaries. By the laws of diffusion, oxygen enters the pulmonary capillaries, where it is taken up by haemoglobin.

There are two sewage systems to cleanse the cytoplasm of metabolic wastes.

The first is the lymphatic vessel system. Its inlet end begins in the body tissues, where the waste products of a large number of cells are discharged into the intercellular fluid. The lymphatic network ends in the right atrium. Through the lymphatic vessels, fluid flows in only one direction - from the tissues to the heart. On this path there are lymph nodes. They contain lymphocytes that neutralise cell metabolic products harmful to the organism, as well as foreign agents.

The second system that contributes to the purification of cell cytoplasm is the venous part of the vascular network. Let me remind you that the cardiovascular system is continuous, blood circulates through it from high-pressure vessels to low-pressure vessels.

- Wait! As far as I know, a capillary is not an impermeable structure, but a porous vessel through which fluid can escape into the extravascular space. Isn't that right?

- You're right! Some of the blood plasma does come out of the capillaries.

- How come our body doesn't swell but our blood pressure doesn't drop all the time?

- With your clever question, you're forcing me to elaborate on some of the anatomical details of microvessel structure.

Each small artery (more precisely, it is called an arteriole) branches many times and gradually passes into a precapillary vessel, capillary, postcapillary, and venule. Then the fusions of many small venules form veins. By merging veins of smaller diameter into a larger vein, they are enlarged up to hollow veins, the merger of which occurs at the entrance to the right atrium. So, in the precapillaries and capillaries blood pressure exceeds the extravascular pressure of the fluid, so the blood plasma partially leaves the vessel into the intercellular space. But along the course of blood pressure gradually decreases, so already in the postcapillaries intravascular pressure is lower than the pressure of fluid in the intercellular space. Due to this, some of the fluid flows back into the vascular network. Normally, this volume should roughly correspond to the percutaneous ejection of blood. Then there will be static blood volume in the cardiovascular system.

- Okay, I get the mechanics of it. But sometimes we swell. What causes that?

- The point is that there are other participants in the process of maintaining cleanliness inside the cells. I haven't said anything about them yet. Well, you should!

- And who are these mysterious participants?

- Excretory organs: kidneys, sweat glands and lungs. They can all reduce blood volume in the cardiovascular system, but to different degrees.

The kidneys are the main excretory organ. At rest, they produce about 1 ml of urine every minute. During breathing, some of the liquid fraction of the blood is released into the air through the pulmonary capillaries in the form of water vapour. In addition to this, we sweat. And the main part of sweat is water in which blood salts are dissolved. In total, these fluid losses amount to about 1.5 - 2.5 litres per day. This is the main reason why we drink water.

- And how much blood does a person have?

- Blood volume is estimated to be about 7% of your body weight. If you weigh 80 kg, your blood volume will be approximately 5.6 litres.

You're taking me sideways! I didn't even answer your question about the causes of oedema. The answer is not as simple as doctors sometimes make it out to be. There's a deep history going back hundreds of millions of years.

- Oh, come on! I ask him a simple question and he talks about hundreds of million years. There were no humans in that antiquity, as far as I know. What could be the connection?

- I'm not surprised at your surprise. But the facts are that the internal chemical composition of our cells strongly resembles the chemical composition of prehistoric seawater. The distant ancestors of land animals evolved in the oceans for a long time. During this time, they developed special mechanisms that monitor the ionic composition of the cytoplasm and intercellular fluid. Hence the famous homeostasis (i.e. maintaining the constancy of the human internal environment in the face of changes in environmental parameters).

I will not go into all the nuances of ensuring this homeostasis, but I will

mention some key points.
So far you and I have used the word pressure exclusively in relation to blood pressure. It is formed by the practical incompressibility of the fluid and the elastic properties of the blood vessels or heart chambers. This was sufficient to understand haemodynamics. When it comes to the processes of mass transfer in tissues and through the cell wall, *osmotic and oncotic* pressures become important components of hydrodynamic forces. Osmotic pressure is caused by the phenomenon of osmosis, i.e. the ability of salt solutions to absorb solvent (in our case water). Oncotic pressure is caused by the property of certain proteins to absorb water.
So, tissue oedema is an increase in the amount of fluid in cells and intercellular space. Edema can be either due to a decrease in blood flow velocity (due to heart weakness) and an increase in intravascular volume and pressure in small vessels, or due to an increase in osmotic and/or oncotic pressures.
I have already mentioned that the kidneys are the main regulator of blood volume in our body. Normally, about 25% of the cardiac output flows through the renal artery. The kidneys consist of two functional parts. In the first one, blood is filtered and primary urine is formed. Its volume is proportional to the blood pressure in the renal artery. In the second functional part - renal tubules is the reverse process: from the primary urine is absorbed into the vascular system about 99% of the fluid. The remainder is the actual urine. Its amount is inversely proportional to the osmotic pressure of the primary urine
(hence blood). Thanks to this organisation of kidney function, the osmotic pressure and ionic composition of the blood remain virtually constant, even though the amount of fluid we take in and its salt content may vary.
As for oncotic pressure, it is determined mainly by the composition of those substances that are absorbed into the blood from the digestive system.
- If I get the gist of this piece, there are three ways to combat oedema: get a good sweat, reduce salt intake and reduce fat intake. Is that right?
- I would add to this by drinking plenty of water. You can speed up the process by using diuretics. But I want to warn you not to overdo it! This will lead to a decrease in blood volume, which will cause a drop in blood pressure and worsen the blood supply to the body.
So, I have outlined the anatomy of the sewer system. In it, the driving force is the pressure difference created by the heart pump.
The human heart consists of four chambers (two atria and two ventricles) equipped with valves. The right atrium and right ventricle pump venous

blood through the pulmonary vasculature. The left atrium and left ventricle pump oxygen-enriched blood into the arteries of the great circulation circle.
The heart is a muscular organ of discrete action. Its work consists of two phases - diastole and systole. In diastole, the heart muscle (myocardium) is relaxed; in systole, the myocardium contracts and develops mechanical force. In the diastole phase, the inlet valves of both ventricles are open and blood flows into the heart cavities under a pressure gradient. As the systole phase begins, the myocardium begins to contract, causing blood to flow backwards from the ventricles to the atria. This causes the ventricular inlet valves to close. At the same time, the ventricular outflow valves are also closed because the pressure in the ventricles at the beginning of systole is inferior to the pressure on the other side of the valves. However, as the myocardium contracts within the right and left ventricles of the heart, blood pressure increases rapidly. Once the intraventricular pressure slightly exceeds the blood pressure on the other side of the valve, the valve opens and the ventricular ejection phase begins. From the right ventricle into the pulmonary artery and from the left ventricle into the aorta. When a significant volume of blood has been expelled from the ventricles, the pressures in the ventricles begin to decrease. At the moment when the pressures in the ventricles are lower than the blood pressures in the corresponding arteries, the outlet valves of the ventricles close. The diastole phase soon begins.
Each cardiac cycle is triggered by an electrical potential produced by specific pacesetter cells localised near the mouth of the right atrium. This area is called the sinus zone. You must remember that the specificity of the cells of this zone is that they have an automatic rhythm of action potential generation. It passes through the conductive system of the myocardium, excites in its cells an electrical impulse, under the influence of which the cell contracts. The actual effect of contraction of the whole myocardium is the result of asynchronous contraction of all myocardial cells.
Note that the cells located between the atria and ventricles also have an automatic rhythm. However, in norm their rhythm is inferior to the rhythm of the cells of the sinus node, so the sinus node is considered to be the main *driver of the rhythm of* cardiac contractions.
Let me remind you that an increase in blood temperature by each degree accelerates the heartbeat by about 10 beats per minute. The rhythm of heartbeats is also influenced by the chemical composition of the blood and the effects of substances secreted in the nerve endings of the fibres descending from the brain along the sympathetic and parasympathetic (vagus) nerves. The sympathetic nerve speeds up the heart, while the

parasympathetic nerve slows it down. Because of this reciprocal nerve influence, the brain is able to regulate the performance of the heart pump. In conjunction with changes in the tone of the arteries and veins, blood pressure is regulated.

- Why regulate blood pressure?
- In my opinion, there is such a deep meaning behind this simple question that it is even scary to say!
- Knowing you, you're not the shy type. So go ahead and tell me why.
- This is going to be a serious and long conversation. Although we have already travelled a certain part of the way today, you will need a fresh head to understand the deeper meaning mentioned. Let's continue this conversation tomorrow. But I ask you to listen to the recording of today's conversation beforehand. It is important for me to know tomorrow how much you have grasped the connection between some of our organs and the physical and chemical state of cells. It is important because I will be building a causal bridge between the physiology of the whole organism and the conditions of life activity of cells.
- All right, I'll try.

CHAPTER 9

Conversation nine

Problems of cellular life support (continued)

- Are there any questions about yesterday's conversation?
- Yes, I do. It's not quite clear to me how an organism that kills tens of millions of cells every hour can sense and react to problems in a single cell.
- The question is legitimate! I'll say more. If you thought that by talking about a cage, I meant literally one cage, that is my omission. Actually, in our conversations, a cage is more of a virtual concept, an image, than a real object. Speaking about a cell, we should understand some virtual object that has basic properties of a cell. At the same time, this object, most likely, should be identified with a large number of cells, localised by fate in the part of the body we are considering. It is convenient for me to work with this virtual object in the same way as, for example, it is convenient for mechanical physicists to operate with the concept of a material point. I hope that in the course of this presentation you will understand more clearly the motives why I use a virtual cell.
- Clarification accepted! Let's move on.
- I have already mentioned that the vast majority of circulatory specialists consider the cardiovascular system as a kind of autonomous object. It has defined internal variables (blood volume, total lumen of vessels, their elasticity) and external functions (pressures in different regions and blood flows).

This paradigm establishes regularities according to which external functions are related to the current values of blood volume, total lumen of vessels, and their elastic properties. The physiology of the cardiovascular system is reduced to the description of dependences of general and regional haemodynamics on changes in the properties of the heart pump and vascular characteristics. Various neurohumoral mechanisms are considered as modulators of these properties.

The main subject of discussion among physiologists is the purpose of the cardiovascular system. There are two points of view: 1) the purpose is to maintain stability of mean arterial pressure; 2) the purpose is to regulate minute blood volume. The first point of view is based on the fact that no blood flow velocity sensors have been found, and all known receptors respond to the stretch created by blood pressure. The second point of view operates on the logic according to which only adequate blood flow can bring cells everything they need for their metabolism and remove its waste

products.
There are compromise schemes in which regulation of blood pressure indirectly leads to regulation of central venous pressure and cardiac pump performance. In these compromise theories, one of the main roles is played by the kidneys as a regulator of mean blood pressure through changes in urine output. There is also a large group of physiologists who believe that the brain has access to regulate the tone of the walls of the smallest vessels of all regions of the body, so as to supply all local needs for blood flow. In my opinion, this view is the weakest. The brain does not have sufficient neural resources to control the smallest vessels. Moreover, there are no required receptors capable of carrying to the brain neurons information about the current state of the body cells.

- What's the problem? As far as I know, blood circulation has been experimentally studied for more than a century. Could it not be determined even in such a period of time what characteristics of the cardiovascular system are target characteristics for its regulators?

- There are problems that are unsolvable, especially when the problem statement is absurd!

- Isn't that a bit harsh!?

- Tough, but to the point!

Let's think back to when there were leaps and bounds in the sciences.
For more than 1500 years, the philosophical picture of the world created by Aristotle and, in part, by Ptolemy, was not doubted by anyone. This philosophy had its own self-sufficient internal logic. Moreover, it was based on observation, that is, on facts. At least in that part of it which concerned the arrangement of the starry sky, the Sun, the Earth and the planets. The schemes of planetary motion devised by Ptolemy worked well enough in the applied navigation of mariners. However, a man came along who questioned whether the Earth was the centre of the universe. Copernicus merely reversed the Earth and the Sun. This reversal was revolutionary for the whole of astronomy. But its significance was, for a long time, merely philosophical. Navigation technologies that used the new conception of the solar system were still more than a century behind the old technologies based on false cosmological ideas in terms of accuracy .
I note this detail because there is a common view that practice is the criterion for the truth of a theory. Or it is quite common to hear in the scientific community that all versions-hypotheses are equally good, only time will show who is right and who is wrong. Truth is not born when it is recognised by the majority. The truth was true even when one saw it and others did not

believe in it.

Let me give an example from the history of medicine. After all, for a long time there was an opinion that blood was created anew in the heart every day. The fact is that when corpses were autopsied, there was no blood in the arteries. It was found only in the veins and in the heart. However, when the English doctor Garvey measured the volume of the heart cavities and multiplied the resulting number by the frequency of heart contractions per day, the number obtained was many times greater than the mass of the human body. It became clear to him that blood must circulate and return to the heart. He realised this truth, despite the fact that at that time capillaries had not yet been discovered. Not only was he cruelly ridiculed. He was shunned by his colleagues. It wasn't until many years later that microscopes proved Garvey right.

Was not the Newtonian doctrine of gravitation and absolute time and space quite logical? After all, it was based on measurements of the trajectories of celestial bodies. However, Einstein with his mental experiments debunked Newton's theory, although he apologised for it.

Did Rutherford see the atom when he proposed his planetary model? Did Niels Bohr know about the existence of stable electron orbits by the time he modified this model of the atom? In fact, long after Bohr, the reason for the stability of the atom was not clear. Only in our time it became clear that the stability of the atom is connected with the exchange of photons between the nucleus and the electron. Only very recently have astrophysicists realised that gravity alone is not enough for stars to assemble into a galaxy. The contribution of electromagnetic fields and dark matter to this process has only recently become clearer, and that is due to calculations on computer models.

- You jump so dashingly between Aristotle, blood circulation and galaxies that even I, a person with a physics and engineering education, who follows the novelties of science, find it difficult to cover everything said. In fact, why such an excursion?

- Only then to convey to you my conviction that any significant change in our understanding of nature begins with the initiator of the new view revising the proposed explanations and going far beyond the dominant paradigm. At the same time, the new paradigm can only claim to be true when all the old facts find their explanation in the new system of views. Here I have brought you to the point where I would like to start my answer to your question about why you need blood pressure.

- What a zigzag! Couldn't it have been simpler?

- Maybe I could have. But, you know, it's a hot topic. I have heard many times from different representatives of science that a hypothesis should be proved by experiments. And how to prove it, if physiologist-experimenter knows how to work only with body fragments (for example, with a separate organ or with several of them) and even in a narrow range of changes in their characteristics. But I am talking about an organism in which a huge number of cells of different specialisation mutually influence each other and form a very complex network. Here there is no other way out but to try to make mental experiments.

- And the computer ones?

- You know that's what I've been doing for over forty years. But all our models, even the most complex ones, are not far removed from physiological experiments. And besides, quantitative modelling requires reliable initial data. Where to get them? And if all the required data were available, then why would there be a model?

- What's the purpose of building models then?

- You know, there is a small group of applied problems when the result is known, some of the vital signs are measured, but there is nowhere to get a couple or three indicators. That's why applied models are built to calculate these missing indicators. Plus, sometimes calculations on models are much cheaper than real measurements.

- Well! So how did you come to your understanding of the role of blood pressure in the body, by mental experimentation?

- In part, yes, but only in part. I had to go through a huge number of scientific publications not only on blood circulation, but also on different sections of biology, to understand energy, biochemistry and cell physiology. On top of all this, I found a way to model organismal phenomena using an extremely simplified - binary - model.

- What the hell is that?

- I will. All in good time.

For now, I'd like to take you back to the basic needs of each of our cells.

As a reminder, regardless of specialisation, each of our cells must constantly solve complex problems. Firstly, to obtain a sufficient variety of starting materials necessary to provide its metabolism, repair damaged structures, and respond to the effects of other cells or other physicochemical entities. Second, depending on the current phase of the cell cycle and on the frequency and power of external influences, the need for an influx of substances varies. Finally, the complexity of the problems to be solved is due to the fact that neither the right assortment of chemicals nor the right amount of chemicals may be available inside the body. The question is: what to do? The answer is to adjust! But this requires coordination of activity of many organs of the body. What mechanisms are capable of such coordination, if I have already said that the brain alone cannot cope even with the fine regulation of local blood circulation?

Answer: through chemical feedback from cells to organs!

- Wait! I think you're contradicting yourself. You said there are no channels to every cell.

- Asserted and asserted! You don't forget that there is a successful "sewer" system consisting of blood vessels and lymphatic vessels. After all, everything that enters this system reaches every cell sooner or later. Of course, the speed of movement of chemical agents, which fulfil the role of information, is much inferior to the speed of impulses in the nerve channels, but there is an advantage: the energy expenditure in the "sewer" system is more modest than that of the central nervous system. One should also not overlook the fact that the chemical agents secreted by cells circulating in the blood are capable of modulating the activity of neurons, wherever they are localised in the chain of nerve-reflex regulation by autonomic systems. It is not for nothing that this dual neurohumoral regulation has long been adopted by evolution.

- But humoral regulation is like shooting at squares! What about accuracy?

- Cellular problems seem to be satisfactorily solved also within the small accuracy that can be achieved by diffusion of chemical agents in the liquid intercellular medium.

I should add that accuracy correlates with the density of the network of microscopic vessels in a given area of the body. This density has been found to be different in different parts of the body. There are special cellular factors (they are referred to by the general term *angiogenesis factors)* that accelerate the formation of new branches of microvessels.

- That's interesting! I heard the opposite from my doctor. He said that at my age, the network of small arterioles is already thinning. Is that true?

- Basically, yes. There is a symptom associated with arterial hypertension. It indicates only one of many scenarios of hypertension development. Nevertheless, angiogenesis, that is, the formation of new small vessels, is observed throughout a person's life. Another thing is the rate of angiogenesis. It depends on many other factors. We will not go into this problem, let specialists deal with it. I would like to return to the issue of the binary model of the organism.

- Is this the model by which you understood the mechanism by which cells

influence the physiology of the body?
- Not only that. I brought my understanding to experimental physiologists. Rethinking of the principles of organism's work was the result of system analysis of one problem, to which nobody paid due attention. I mean the problem of the balance of energy production and consumption in each cell.
- ATP molecules?
- Yes, to make a long story short. In fact, we will only talk about the average rate of aerobic synthesis of AMP, ADP and ATP in the mitochondria of the cell.
- But I remember that we have touched upon this topic before and you said that the problem of energy balance has been solved even in single-celled organisms like yeast. Why did it arise again in our cells? After all, you also said that our cells have inherited autonomous mechanisms to ensure energy balance in them when the rate of ATP consumption changes. Didn't you?
- Now that you remember all that, it's easier for me to explain the nature of the new problem.
Autonomous mechanisms are able to overcome ATP deficiency only at its moderate value. When the energy demand increases sharply and significantly, these mechanisms become ineffective. A state of energy deficit develops in the cell. Without energy it is impossible to carry out biochemical transformations to the end, nor to maintain optimal ionic homeostasis of cytoplasm. Such a cell falls into a state of oppression. If the cell was part of the functional systems of the organism, the quality of their functioning deteriorates. In extreme cases, when the lack of energy covers a large number of cells, the organism becomes lethargic, sick. It is clear that such an organism has little chance of survival in a competitive environment. But there is a way out!
When biochemical transformations are interrupted at some intermediate stage, among the many intermediate chemicals are those that act as negative feedback signals and activate mechanisms that ultimately accelerate aerobic ATP synthesis.
Currently, biochemists have discovered more than ten chemical substances of this type. Some of them accelerate the process of mitochondrial fission, others promote the growth of their size. Both named changes work towards the common goal of increasing the aerobic capacity of ATP production. However, there is one important detail here: in order to increase the total area of the cell's mitochondria, it is necessary to accelerate the influx into the cell of all those ingredients from which mitochondria are synthesised.
- And how did the body solve this problem?

- Rather elegantly: he used another part of the intermediate products of interrupted biochemical transformations, namely those small molecules that leaked through the membrane of the depressed cell and out into the intercellular fluid.
- I'm already guessing what might happen next. They get into the bloodstream and with the blood flow they reach the organs that produce the building blocks for mitochondrial synthesis. Isn't that right?
- You're becoming more and more of a physiologist! Ten days ago I wouldn't have dreamed of it.
- So my brain isn't so old and stiff. I'm still able to learn new things! I'll pass your appraisal on to my boss. He should give me a raise!
- Stop dreaming! Let me tell you in general terms how the organism promotes an increase in the rate of ATP synthesis in those cells where it is necessary. Let me say right away that evolution has accumulated several independent ways of solving this task in our body.
- Wasn't it because the role of energy for the body was so important?
- I think so. But there are other, perhaps accidental, preconditions for diversification of pathways and ways of energy supply to cells.
- Takeaway. "Energy diversification" is almost the most frequent phrase heard in the EU in recent years. I have learnt one thing from politically motivated mass media sources: having many ways to achieve a goal is a guarantee of reliable survival.
- The analogy is apt! Many paths are better than one.
- Wait! Another thought occurred to me. Is it necessary to increase the total area of mitochondria in the cell? Are there no other ways to increase the aerobic capacity of ATP synthesis with the existing mitochondrial area?
- I didn't expect such a subtle question from you! You're on a roll today! I'll answer. I do. There's more than one.
Firstly, there are special chemicals in the mitochondria themselves, the concentration of which affects the rate of ATP synthesis.
Secondly, it is known that the basic regulator of the rate of ATP synthesis is AMP concentration. There is a chain of regulation of AMP concentration by increasing blood glucose concentration.
Third, mitochondrial productivity can be increased by increasing blood flow to the affected cells. For this purpose, there are local vasodilators that are released from the depressed cells.
Fourthly, it is possible to increase the concentration of oxygen in the incoming blood. This can be achieved by increasing the frequency and/or depth of breathing (i.e., pulmonary ventilation) and accelerating red blood

cell production (erythropoiesis). Finally, blood pressure can be increased. There are cardiac, vascular, and a third, total volume increase, pathways for this.

By the way, one of the best known cellular factors regulating aerobic ATP synthesis is the so-called *hypoxia-inducible factor*. It promotes the production of erythropoietin, a protein that regulates erythropoiesis.

- If I remember our conversation a couple of years ago correctly, you were the one who discovered the energetic mechanism of arterial hypertension.
- That's saying it loudly! I'm being logical. A discovery is when an idea is proven experimentally. Unfortunately, no such purposeful experiment has been set up so far. On the other hand, the existence of indirect data supporting my hypothesis makes such an experiment unnecessary.
- You've listed many ways to solve the problem of getting a cell out of a depressed state. Can't they all be engaged at the same time?
- Why not? In the real organism, they often overlap. Moreover, each person may have his or her own preferred scenario for dealing with energy deficiency.
- I get it. You mean ontogenetic adaptation.
- Exactly!
- I thought I understood everything. But you still haven't said anything about the binary model. Will I wait?
- We're already standing next to her!

Each cell has its own dynamic energy status. The binary model assumes that those cells that have enough energy to perform all kinds of biological work are satisfied with their state and do not seek to change the state of other cells and organs. And those cells, which experience energy deficiency, influence by products of their metabolism on life-supporting organs, increasing their share of energy raw material inflow. In other words, we have divided cells into two types: satisfied and unsatisfied.

- Elegant! But doesn't this categorisation of cells seem unnecessarily crude? Is everything taken into account in this approach to their coexistence?
- Certainly not all of them! At least the specific contributions of those mechanisms that work on the chemical homeostasis of the cell cytoplasm. We have already touched upon this problem in passing. I will now go into a little more detail.

In my reasoning it was tacitly assumed that a priori there is a certain basic level of activity of the organism. Physiologists have long ago agreed to take the state of a person at rest as the basic one. All changes caused by ATP deficiency in cells increase physiological activity of the organism. It is also believed that in a state of any statics, all processes in cells are balanced regardless of metabolic rate. Theoretically, such an organism can exist for any length of time.

- However, in discussing the problem of canalisation, we have already touched on another aspect of cell life: the cytoplasm must be in perfect condition. How does the binary model of a cell community account for the problems caused by cytoplasmic pollution?

- There you go again! Let me just say that you're not the only one asking this question. I had some time to think about the answer. And I found it: it was first published in 2017 in my English-language book, which in Russian translation will be called "Optimal Circulation".

So, about insufficient quality cytoplasm.

We have already said that a cell is an optimal chemical machine when the physicochemical composition of the cytoplasm is optimal. There are two components of the required optimality - biophysical and biochemical. The biophysical concerns the ionic composition. Since the major ions are potassium, sodium, calcium, chlorine and magnesium, their concentrations must be such that the resting potential of the cell is normal. This is especially important for excitable cells.

- I remember. These cells make up our functional systems.

- That's right! Let's move on.

The biochemical component of cytoplasm optimality is determined by the presence of a sufficient variety and proper concentrations of primary ingredients, as well as a minimum concentration of metabolic wastes. Note that the number of primary ingredients includes those of them, with the help of which ATP molecules are synthesised. That is, we believe that the energy problem is a private problem of optimality of the chemical composition of the cytoplasm. The main products of cell metabolism are carbon dioxide, hydrogen ions and OH-.

We have already mentioned that when arterial blood passes through the kidneys, it is rectified (i.e. excess potassium, sodium, calcium, chlorine and magnesium ions are removed). In other words, the kidneys are the main organ maintaining ionic homeostasis of the cell cytoplasm. The sweat glands perform a similar function.

Now let's talk about how excess carbon dioxide is removed from the blood. You already know that it happens when venous blood passes through the lungs. I would only add that the intensity of carbon dioxide removal increases with increasing pulmonary ventilation and blood flow. Hence, there must be some organ, by necessity, regulating these two physiological variables.

First, let me tell you how the body knows that the need has arisen. In the area of the aortic arch and near the branching point of the common carotid artery into the internal and external carotid artery, there are special chemoreceptors. They are sensitive to the acidity pH of the blood and to the concentration of CO_2. An increase in these blood characteristics activates brain neurons that increase pulmonary ventilation and blood pressure.

- So the same reflex controls the activity of both the lungs and the cardiovascular system?

- Yeah.

- I understood that increased pulmonary ventilation accelerates the clearance of CO_2 from the blood. Why would you raise blood pressure?

- The increase in blood pressure is mainly due to the increased pumping function of the heart. This leads to greater blood flow through the lungs and accelerates the replacement of CO_2 by 0_2.

- Okay. But then it turns out that instead of a binary model, it would be better to create a model that takes into account the four states of cellular status.

- Good for you! I'm working on it. As you have ascertained, the physiology of energy balance in our cells and optimisation of their cytoplasm is clear in general terms.

That's it, I'm tired, even though you look like a cucumber! We'll continue tomorrow.

CHAPTER 10

Conversation ten

The great compromise of cells

- Any questions about last night's conversation?
- No, I don't.
- Moving on. Today I'm going to try to convince you that the state that medicine calls health is actually a compromise between all of our cells.
- Are they making a deal?
- Naturally, they do not know how to do this. And compromise is always necessary when the interests of the participants in the process do not coincide in everything. That's where we'll start.

Every cell in the body is preoccupied with its own problems. Roughly speaking, it needs to eat, get rid of waste products, reproduce from time to time, and defend itself against destructive influences, wherever they come from.

As a rule, there are fewer resources of total consumption in an organism than can be assimilated by all the cells of the body. Therefore, the egoistic aspirations of each cell to maximise its consumption are limited by the similar aspirations of others. Unsatisfied cells change the physiological state of the body in such a way that in a network of trillions of cells, without any "directors" and "rulers", a compromise distribution of common resources is obtained. This is the optimal state under the existing resource constraints.

- Reminiscent of Nash equilibrium in game problems and economics.
- The analogy is a beautiful one!

But there are a number of nuances in the body.

Let's talk first about how evolutionary mechanisms have created preconditions for deviation from the mentioned optimum.

We have already mentioned that the main way of redistributing resources in favour of an organ is to increase its blood flow. With the proper variety of nutrients in the blood, the more it flows to an organ, the more of these nutrients it will receive.

- If only the cells of the organ are able to assimilate these resources!
- You're right. My statement does not work if the rate of assimilation of chemical ingredients is lower than the rate of ingestion: the excess substances will leave with the venous blood.
- What about stocking up?
- Yes, there are some such mechanisms. But they are able to store only certain resources. For example, energy resources, the supply of which from

the outside is not constant, and without energy there is nowhere, are stored in two main forms. Part of excess glucose is converted into glycogen and stored in the liver and muscles, while the other part of this excess is converted into fat and stored in fat cells.
But oxygen, which is the second major consumable ingredient of oxidative phosphorylation reactions in mitochondria, is not stored.
- And why stockpile it when it's always available - all you have to do is breathe it in!
- On the whole, of course, you're right. The evolutionary benefits of oxygen storage don't seem to be apparent. But this is true if an organism (aquatic or terrestrial) does not change its habitat. And there are some that more or less regularly change from one environment to another. For example, secondary aquatic animals (cetaceans) or those species that get their food in the water, and on land rest (crocodiles, sea snakes, iguanas, seals, elephants, calans and leopards). All of them continue to breathe with their lungs, even though in the embryonic stage of development they also develop gills, which are then eliminated.
- As far as I know, these animals have specific mechanisms that allow them to significantly extend their time underwater.
- Yes, a number of adaptive adaptations are described.
But I've taken you away from the main point. Let's get back to the man.
The nutrient and energy requirements of different organs can change significantly from time to time. These changes are usually caused by a factor external to the cell, such as the brain forcing it to participate in a behavioural act (running or walking, fighting). In order that an organ whose consumer demands have increased may fulfil them, its arteries must be dilated. If the demands are so great that at the current level of arterial pressure, simply by dilating local arteries, it is impossible to fully satisfy the needs of the organ, it is necessary to raise the arterial pressure. This is possible by narrowing the arteries of those organs that do not need much blood at the moment. Another alternative is to increase the pumping activity of the heart.
- I remember that we considered these possibilities when we talked about mechanisms for restoring energy balance in cells.
- That's good to hear! So a common understanding is already in place.
But I want to turn your attention to the fact that "cellular communism" in the evolution of our organism, and not only our organism, did not take shape.
- What, are you implying that the body has its own oligarchs or something?
- Maybe the word "oligarch" is not the best definition in our case, but it is a fact that there is no equality of organs. At least two organs, the brain and the

heart, have escaped the general principle that they should share with the organ in need. The arteries of the brain and heart do not submit to vasoconstriction. Even under conditions of severe deficiency of energy and oxygen, the vessels of these organs are not constricted by the central nervous system. The blood supply to the heart and brain is ensured by a substantial increase in arterial pressure. But such emergency mode of blood circulation cannot last long. In the organs whose interests are "pinched", intermediate agents of interrupted metabolism accumulate, their vessels dilate and blood pressure drops.

- Does it cause fainting?

- Theoretically, yes. But fainting, or circulatory shock as it is called by experts, is more often caused by pain shock. In severe pain, the brain dramatically dilates peripheral blood vessels, and blood pressure instantly drops to levels where there is virtually no blood flow to the brain. Syncope causes the person to fall. As a consequence, the hydrostatic column of blood between the heart and head disappears, cerebral blood supply conditions improve and the person soon regains consciousness.

- This is some kind of extreme circulatory regime. And how, in everyday life, when the nutrient needs of trillions of cells fluctuate rapidly and significantly, does the body manage to meet those needs? Is our minds constantly preoccupied with this physiology?

- This is a very significant, if not the most important issue of integrative human physiology. In previous conversations we touched upon the private sides of this global problem. I will try to answer this question briefly.

Let us imagine an ideal situation of sensory deprivation, i.e. the brain receives no impulses from either internal receptors or external sensory organs. In this hypothetical situation the functional integrity of the organism and all its cells is preserved only if a certain minimum frequency of impulses to all organs is sent through the nerve channels descending from the brain. These impulses have a special - tonic - role. They provide the basal tone of all vessels. More precisely, the basal tone is formed from the joint action of nervous and humoral influences on vessels. After all, each of our specialised cells continues to live, i.e. to consume resources and release chemical substances - metabolic products - into the general intercellular environment.

Let me remind you that cells can be in different phases of the cell cycle, each phase has its own metabolic rate. I will also note that the cell autonomously fights against all spontaneously acting destructive forces. First of all, these forces disturb the ionic homeostasis of the cytoplasm. Its restoration requires energy. The total energy demand of all body cells in the mode of their minimum integral activity is the energy equivalent of the physiological concept of the *basal mode.* I will note that physiologists themselves do not define this mode as strictly as I have done now. They do not require general sensory deprivation. In practice, the activity of the organism at rest when the body is horizontal is taken as the basal level of physiology. To be even more precise, this state is described only by a limited set of measured physiological characteristics (body temperature, blood pressure, heart rate, respiration).

Now let us imagine that signals are received in the brain via external sensory channels that the physical and chemical characteristics of the environment have changed. These impulses reach certain groups of neurons in the cortex of the large hemispheres. Some ensembles of neurons go into an excited state. With the dynamics of external flows of impulses, the expression of this state changes. It can be multimodal (when impulses come from different sense organs) or unimodal (when impulses come only from one sense organ). In the first case, the maximum number of synchronously excited cortical neurons can reach tens or even hundreds of millions. In fact, this is the biophysical basis and correlate of consciousness. We perceive any changes in the number of synchronously activated neurons as changes in the brightness of consciousness. This is how a multimodal model of the environment is formed in our brain. It is rarely stable. With fluctuations in the intensity of flows of incoming information, consciousness flickers, becoming brighter or dimmer.

- Aren't you saying that attention span is somehow related to an increase in the number of synchronously activated neurons?

- I think you're right.

- Why do we sometimes find it difficult to focus for long periods of time on one thing?

- I think not the least of these is the difficulty of maintaining the required level of blood supply to all the neurons involved in a given intellectual activity.

- I once watched a science programme about how neuroscientists and psychologists are trying to objectify the thinking process using CT scanners. I read scientific articles about attempts to control technical devices with thought. And I noticed how the colours of different areas of the brain on a CT scan changed very quickly depending on what activity the subjects were doing. Something like the flickering of neuronal ensembles was indeed visible.

- I see what you're talking about. It's called fMRI - functional magnetic resonance imaging. It differs from conventional tomography in that it

registers rapid changes in blood flow in fairly small areas of the brain. It is believed that the blood flow value correlates with the functional activity of a local group of neurons. The resolution of the method is hundreds of times higher than that of electroencephalography. Therefore, fMRI is increasingly used in scientific and diagnostic research.
In my opinion, it is unlikely to expect commercial application of fMRI for control tasks of complex technical devices and systems. Most likely, this experience will be useful in the next stage of development of the technique of recording human thought activity.
Let's get back to the main topic.
Let me remind you that we also talked about the fact that there are nerve outgrowths running from the cerebral cortex towards its more ancient parts. These impulses reach those neurons that function in local subconscious networks as regulators of internal organs and systems. Even in sleep, when vision is switched off, but hearing, tactile sense and olfaction work, activation of these sensors leads to modulations of activity of organs of vegetative systems: indicators of cardiovascular system, respiration, thermoregulation, organs of internal and external secretion change. In dreams, these organs imitate practically the same modes of functioning that would take place when awake.
- Why don't we walk or run in our sleep?
- Evolution has eliminated such individuals. Normally, when you fall asleep, the path of impulses from cortical neurons to motor neurons is blocked. Or rather, it's severely weakened. Small movements remain. Moreover, where it was not associated with a risk to life, these channels function. A striking example of this is the movement of the muscles of the eyeballs: when we dream about something visual, the eyeballs move in the same way as they do when we are awake. I won't talk about sleepwalkers. I will only limit myself to the fact that evolution seems to have shown a specific sense of humour here.....
- You have a peculiar sense of humour! People are in grief, and you're making jokes....
- All right, listen up.
When the brain (I call it the "despot") sends its stimulating or suppressing impulses to the various organs, which is the only way to integrate the organs into a single functional act, it, the brain, does not ask these organs whether they want to live in the imposed metabolic regime or not. The brain does not even care to properly increase the blood supply to the stimulated organs. And in fact, the cells have a very meagre supply of ATP molecules. They are

sufficient only for a very short time.
- But we don't fall down after taking a couple of steps! So, the brain somehow builds up the supply of working muscles and other associated organs. Isn't it?
- I don't think I agree with you. Initially, under these conditions, the brain sends vasoconstrictor impulses to almost all organs to increase blood pressure. The rise in blood pressure is necessary to maximise muscle output in the name of the brain-imposed organismal goal ("Run!" or "Fight!").
- Either I'm missing something or you're saying the wrong thing. An organ can't work if its blood supply is impaired!
- Yes, but there are nuances in different agencies.
Take muscles, for example. They are adapted to contract for quite a long time and without good blood circulation and oxygen and energy supply. Firstly, as it was mentioned earlier, muscles have glycogen reserves - energy raw material. Secondly, a muscle cell can effectively use anaerobic way of ATP synthesis. It is true that lactic acid accumulates in the muscle. But if its concentration is not prohibitive, you can work. And afterwards, if you are far away from predators and you are not eaten by them, you can take a breath and slowly utilise the accumulated lactic acid. It is important to remember that it is not the brain that provides this utilisation. Lactic acid itself and its breakdown products have a local vasodilatory effect, increasing the flow of oxygen with blood. And oxygen favours the breakdown of lactic acid into water and carbon dioxide.
In much the same way, local vasodilatory agents provide escape from the oppressive efforts of the brain, restoring cell metabolism and long-term survival.
- But it can't be considered a perfect mechanism!
- And evolution has never created perfect biological mechanisms. Why, one might ask, would a potential prey have a perfect mechanism for evading a predator if the predator is not perfect? The predator's body is made up of the same imperfect elements-cells, and its brain also does not care about the long-term well-being of its cells. The predator generally knows its capabilities, so it uses its brain to surreptitiously get as close to the victim as possible. And the victim knows what the predator is capable of, so it tries to keep a safe distance. The result is two imperfect organisms engaged in an eternal struggle for existence.
- And what, there's no exception to that rule?
- I do. And perhaps this seems to me to be one of the rare but striking manifestations of the advantages of advanced intelligence.

I remember watching a programme on the Discovery Channel on TV once. It was either Animal World or Civilisation. I can't remember exactly. But I do remember the explorer, aka the host, was on a hunting trip with two little half-naked Bushmen. It was a clear sky. The guide was talking about the terrible heat. They walked for a long time until they saw some beautiful forked antelope grazing alone. When it saw the people, it ran away. The Bushmen followed. The speed of the antelope was much faster than that of the Bushmen. The host, looking at this, hastened to say that the Bushmen had no chance of a successful hunt. This he also told the hunters. They only smirked and asked him to follow them. The leading man tried not to lag behind. The Bushmen pursued the antelope persistently, but without any pressure. After a while the antelope became exhausted and lay down. Breathing heavily, it only looked at the approaching Bushmen, who caught up with it and calmly thrust their spears into the antelope.

I remembered the amazement and exclamations of the presenter for a long time. I exclaimed: "Here indeed, knowledge is power!" The experience of the weak Bushmen took precedence over the physical strength of the beautifully built antelope, which had escaped many times from the fearsome predators of the African savannah.

- Look, this already sounds like a parable about evolution and intelligence!

- You can't argue with that!

- But I've noticed that you're not very complimentary about evolution. Why is that?

- It would be more accurate to say that I have a sober attitude towards evolution. I recognise that evolution has produced many great solutions in organisms. Modern bionic engineers should take advantage of these biological prototypes. At the same time, I know that evolution has accumulated many solutions that, if designed by an engineer, would certainly get him fired from his job for professional unfitness. Eyes, for example, are certainly a great evolutionary achievement. But even in this example, by placing the blood vessels of the retina in front of the light-sensitive cones and sticks, evolution has reduced their efficiency.

CHAPTER 11

Talk eleven

Why do we get sick?

- What questions?

- I have, in general, understood the essence of health when viewed from the perspective of cells. But it does not yet follow from this understanding that diseases are inevitable. Can you at least tell me why health leaves us as we age?

- I don't think I can answer this question exhaustively. But I will try to bring some clarity to this problem.

First of all, let's discard those diseases that are the result of injuries. It is clear why: injuries are accidental.

Secondly, let's also exclude infectious diseases. Although there are many of them, as a rule, they affect people with weak immunity. We have only touched on the immune system in passing, but even that is enough to say: this system is weakening due to insufficient reproduction of specialised cells. There are many immunosuppressors in today's technological life.

Let us concentrate on a special class of diseases, the characteristic feature of which is that they develop as if stealthily, so that for a long time we do not pay attention to them.

- If I guessed your train of thought, you want to talk about diseases of adaptation.

- Some call them diseases of adaptation, others call them maladaptation. The essence of this does not change: it means that the body's adaptation mechanisms are no longer able to fully compensate for the negative effects of internal and/or external stress factors.

We have already said that more or less complex biological molecules are usually unstable. Biochemists explain this by the fact that they are based on weak bonds of tertiary and quaternary structures. Many perturbations of even small magnitude (such as those caused by thermodynamic fluctuations of electrons and protons), can break these bonds. Such events in our body are so frequent that about 40% of the energy of ATP molecules is used for re-synthesis of disintegrated macromolecules. It follows that any stresses and problems that reduce the rate of ATP synthesis in the cell inevitably worsen its functional state.

In his time Hans Sellier, the author of the concept of general adaptation syndrome and stress, gave the first theoretical explanation of the causes of a vast class of diseases. But he connected the concept of stress mainly with mental illnesses, explaining the mechanism of their psychosomatic manifestations, which, in my opinion, slightly distorted and narrowed the very concept of stress. And nowadays the term "stress" is more often used as the most general characteristic of the first stage of the process of psychosomatic adaptation of a person to changed environmental conditions. In my opinion, such a one-sided approach to the phenomenon of stress excessively exaggerates the role of the central nervous system in it. At the same time, the contribution of intracellular mechanisms in the struggle for optimal metabolism is undeservedly diminished.

It should be added that the traditional understanding of the etiology of diseases due to stress does not take into account chemical feedbacks through which cells modulate the work of organs and systems, including activation or suppression of the central nervous system itself. Moreover, in chronic altered states of the organism, it is due to humoral communication channels and various chemical agents that cellular rearrangements occur. The changes in the number of cells adapt the organs to the current conditions in such a way as to achieve maximum cellular well-being with minimum energy expenditure.

The first thing that follows from this understanding of the organism's self-adjustment is that we are healthy as long as there is timely adequate adaptation to the changed conditions of life. Naturally, the transition process from the previous quasi-stable state to the new one cannot be instantaneous. The duration of the process largely depends on the vastness of the problem areas. In other words, on the number of cells in which there is a deficit of energy and/or building materials, or the required composition of cytoplasm and/or its ionic homeostasis is disturbed. This is one aspect of the problem. It stems from the fact that overcoming the problems encountered is delayed as their magnitude increases. But there is another, equally significant aspect. It is also fundamental and is an initial flaw, figuratively speaking, a "birthmark" of the mechanism underlying all cellular biosynthesis processes. I am referring to the unreliable apparatus of DNA reproduction.

Since point gene mutations often occur with each act of RNA copying, they accumulate with age.

This often results in an increasing number of erroneous versions of newly synthesised molecules. Immune mechanisms detect, tag and eliminate them. The required molecule has to be re-synthesised. And this is a costly process. In other words, the efficiency of biosynthesis decreases with age. To maintain the performance of the same number of cells with age we have to spend more energy. This is the main reason for the increased vulnerability of the elderly to all affecting factors. An additional reason is that the total number of cells in each specialised population decreases. This is probably caused by the shortening of telomere length as it approaches the Hayflick limit.

- That sounds fatalistic! Are there ways to deal with telomere length shortening or mutations?

- I would separate the two questions.

Let us first discuss the problem of telomeres. In the last twenty years, there has been a lot of research aimed at finding out whether the natural process of telomerase synthesis can be stimulated in some way. Let me remind you that in some cell types this enzyme restores shortened telomeres. These studies echo others that attempt to understand how and why passive parts of the genome switch to an active state and vice versa. This trend goes by the general name of "epigenetics". What I have read on epigenetics suggests that physical activity - walking, running - enhances telomerase synthesis.

- Oh, so I'll be rewarded! There's a reason I love walking and cycling. I'll age slower!

- If I'm alive, I'll be happy for your longevity!

And meanwhile let's tell a little about another hypothesis connected with the reason of age increase of mutations. Different hypotheses about the causes of mutagenesis were many. According to each hypothesis proposed and ways to combat this undesirable phenomenon. I think you have heard of the damaging effect of reactive oxygen species and the means of defence against them. Another name for this phenomenon is oxidative stress. Combating the unwanted effects of oxidative stress has long been big business. However, there is no firm scientific evidence that it is reactive oxygen species that are to blame for the increased frequency of mutations.

- Why do almost half of adults have arterial hypertension, diabetes mellitus and other non-infectious diseases? Why does it happen that you have been healthy for a long time, felt nothing bad and suddenly a preventive visit to the doctor reveals that you are already a patient? Moreover, the doctor says that your disease is chronic and cannot be radically treated. Your only salvation is to take a lot of pills regularly for the rest of your life.

- I won't remind you that life is a dangerous phenomenon, people die from it. You have to be philosophical about everything. There is no immortality. And why should there be? Would you really want to live forever? Personally, I've never wanted that. I've had enough of one life. I've learnt what it's all about. I don't need more.

- Well, I wouldn't mind living longer and disease-free. It's the prospect of

senility that scares me.
- And have you never been intimidated by the incongruity of intelligence with the way of sustaining life that dominates the animal kingdom?
- If you mean devouring other creatures, that's disgusting to me. But futurologists paint a picture of a future where intelligent beings extract energy and chemical ingredients for self-sustenance directly from solar energy and the environment. A kind of intelligent plant with the ability to move freely in space. How's that for an idea?
- In my opinion, the creator, if there was one, should have created such thinking beings. Otherwise we are some chimera with the rudiments of morality and thinking. And the absolute majority of people consider the purpose of life to be to eat more, get bodily pleasures and exert themselves less.
- Let's not engage in moralising. Rather, what do you think about the causes of the so-called diseases of old age?
- I thought I had already answered this question in general terms: evolution has not created and does not create a perfect organism. We are burdened with various mechanisms that only give us a chance to pass on a part of our genes with all their imperfections to the next generation. If an individual manages to do this and the environmental conditions are favourable for the heirs to grow up, reach puberty and also pass on their genes to the next generation, then the species exists. In old age we no longer reproduce, so a prosperous old age is not one of the criteria by which organisms pass through the sieve of evolution.
- Either you are so pessimistic today, or life itself is like that. But I wouldn't want such an unhappy picture of the most grandiose phenomenon of the universe - life - to emerge near the end of our conversations.....
- I might agree with your assessment of life. Indeed, as thinking beings, we recognise life as the greatest evolutionary achievement of the universe. But that's only one side of the perception of life. With the development of genetics came another realisation: we have many flaws in us! That's life! The question arises: what to do?
- I think the answer is unambiguous - to correct nature's mistakes!
- That's the techie in you. I guess I'm on your side, too. But with some reservations.
Firstly, to get rid of accumulated unfavourable mutations once and for all, we need genome correction at the early stages of embryo development, even before gastrulation.
Secondly, experts do not yet fully understand how our genome functions.

Replacing mutated genes with the correct ones, and such experiments have been conducted on animals, sometimes leads to unexpected results. Consequently, the time has not yet come to intervene decisively in the human genome.

Thirdly, we live in a society where the worldview of most people is shrouded in religious beliefs about life and the exceptionalism of man in particular. These views have shaped public morality, and laws are passed under its influence. Today, as far as I know, no country in the world allows the manipulation of human genes for the purpose of improving the species. There are some cases when point mutations in some chromosomes of a sick person are corrected for medical reasons.

- It seems to me that the state of Israel is ahead of the world in this respect. After all, there is a law there, supported by the rabbis, that couples about to marry must undergo genetic control. If they carry certain genes, the marriage will neither be approved by the rabbis nor registered by the state.

- Yes, I'm aware of that. But Israel is the exception so far. For the Jews, reason has trumped religion in this matter. For too long, this nation, forced to live without a state of its own but stubbornly unwilling to assimilate, has practised close blood marriages. The number of mutant genes in the genome of European Ashkenazi Jews is quite high. And this increases the probability that inactive (recessive) genes of the father and mother can form an active mutant gene. Most often, such genes contribute to anomalies.

- Are anomalies always ugly?

- I wouldn't put it so crudely. There are anomalies that impair thc quality of life and require special medical care for carriers of such genes. This is not only a medical problem, but also an ethical problem. But there are other kind of anomalies - the emergence of some kind of superpowers. Mankind is proud of geniuses, and Jews are particularly rich in them....

- And why don't all small and isolated peoples have the same frequency of geniuses?

- In my opinion, genes alone do not explain everything. A lot depends on what qualities are cultivated by a community. Some peoples honour strength and the ability to endure the hardships of life, while others do not. Perhaps the abnormally high number of Jewish Nobel Prize winners is also due to the fact that knowledge is honoured in this nation. Parents from an early age contribute to the manifestation of genetic talents of their children. The fact that there are many Jews in countries where science is developed also has an effect. After all, to become a Nobel laureate, it is not enough to have genius. You also need an advanced scientific environment. But this is just my subjective opinion on such a delicate issue as the relationship between genes and intellectual, physical and mental characteristics of an individual. Statistics show one more thing: the line between genius and mental illness, such as schizophrenia, is too thin.

- I see you have turned the conversation back to diseases. Should I conclude that the diseases of old age are genetically predetermined? If so, why do they manifest with such a delay? After all, I thought that the bad gene manifests itself from the first days of life!

- Indeed, this is a difficult question. I think that the correct answer should not exclude both options: there are genetic diseases that manifest themselves immediately, and there are those that can appear with age. In the latter case, most likely, the activators of "silent" bad genes are those factors of life that act through epigenetic mechanisms.

- There's that mysterious epigenetics again!

- Yes, it is still largely mysterious. But there is already convincing scientific evidence that under certain conditions epigenetic modification of parental genomes can be transmitted to the next generation.

- Is this the new Lamarckism?

- Almost. Perhaps old Lamarck was right about something after all. Otherwise, how else can we explain that sea snakes have a body and tail flattened like fish, all secondary aquatic animals have webbing between their toes, and in the cetacean family the dorsal fin appeared? This list of adaptations is much longer than the examples given. And the assumption that random mutations and selection have produced such structural features because they are effective is hard to fit into a limited time period. Perhaps some activators of epigenetic mechanisms activate a group of linked genes at the same time ...

Perhaps, in this connection it is necessary to remind about another curiosity of genetics. After all, the founding fathers of this science and most of their students did not even think of any other way of inheriting traits than strictly by genes from parents to offspring. However, later it turned out that there is another - horizontal or interspecific gene transfer. By the way, this method is used by microbes. It may seem funny to us, but microbes also have male and female sex. A representative of the male sex has a peculiar sexual pyle, through which the donor cell transfers genetic material to the recipient cell (it is important to note, not only of its own species!). Today, horizontal gene transfer by means of special phages is used in genome correction.

- It's scary! Is it possible that if we get a viral infection at a young age, we can pass on foreign genes to our offspring through natural reproduction?

- In principle, there is. But it should not be feared so much. Geneticists have established that more than 15% of the human genome has been introduced by viruses. In other words, some part of the genome of viruses with which we had contact has not disappeared without a trace. Many viruses have a special adaptation that allows them to insert themselves into the genome of the host cell. Viruses left some of their genes in our genome and gave us new qualities. We should be grateful to them.
- Wow! It turns out that we owe not only to gut microbes, the microbe that swallowed the ancestor of the mitochondrion, but also to viruses! So much for the unexpected zigzags of evolution! It's a bit scary. I mean, none of this could have happened, could it?
- There's more to come!
- What else are you gonna scare me with?
- Not to scare you. It's just that all of the past does not guarantee that our bodies and the human species will stand up to fundamentally new threats. Our immune system protects us only from those pests which it has either met in the past, or in a new encounter it has enough time to overpower the viruses before they become too numerous. The immune experience of the human population is not unlimited. There is no guarantee that the rate at which a virus takes over the human genetic apparatus will never exceed the rate at which an effective immune defence develops.
I note once again that life, no matter in what superlative degrees of comparison we speak of it, is not without flaws. It is not only the individual who is mortal. Life itself is mortal. At least on planet Earth: because inevitable changes in the physics of the Sun will sooner or later deprive life of comfortable conditions.
- What are we supposed to do? We are not ready to improve, and we are not allowed to.
- Let's hope there will be no such viruses for quite some time. Another strategy involves the spread of terrestrial life forms to other planets. But even if humanity manages to implement this strategy, alien life will undergo such an evolution that our distant descendants will not resemble us at all. Of course, this is no reason not to work in this direction. To say more, it is our duty to our not so reasonable genetic ancestors and relatives. In a cosmic catastrophe, they have no chance of survival at all. If our intelligence could save unintelligent life from extinction, I would consider it not only a noble but also a worthy goal of reason.
- What else can you tell me about health?
- Actually, this topic is inexhaustible and multifaceted. When I started our

conversations, I did not set myself the task of covering the vastness. My goal was much more modest - to give you a basic understanding of our body, based on the knowledge of its cellular structure and on the evolutionary worldview. I think everything that could be done in such a small amount of time has been done. We had active discussions. If desired and possible, I am ready to continue and develop the main theses of our discussions in the future. We will take a break tomorrow and meet again in a day. I would ask you to go through all the notes, and during the last conversation to summarise your understanding of the essence of life.

- I'm not going to dodge the exam, and I hope I pass it. After all, during our regular conversations I have already listened to the recordings many times. But considering my boss's interest in everything that concerns a man, and especially his intellectual activity, about which nothing has been said, I have another suggestion. Let's meet tomorrow, too. The topic of conversation is the brain.

- Actually, that's not my speciality. I have always avoided extending any purely physiological regularities to the psycho-emotional sphere. Not for nothing did I mention Nobel laureate Crick's admission that the brain remains largely a dark area for science. But, since you ask, I'll try.

CHAPTER 12

Conversation twelve

Brain

- The very attempt to explain the rules of the brain in one short conversation cannot be regarded as anything other than a gamble. Keep in mind that I am doing this purely out of compulsion and old friendship.
- Don't be flirtatious! I think you know a lot more about the brain than those experts who have been trying hard to model intellectual activity for a long time. They have the nerve to use the terms "genetic algorithms", "neural networks", which have a very distant relation to these concepts in biology. Didn't they call primitive algorithms of pattern recognition artificial intelligence?
- It's not just my limited knowledge that's the problem. You're asking me to almost conclude whether the human brain is perfect, or whether it has the same evolutionary flaws as the rest of the body. Although I have subjective thoughts on the subject, I can't make that judgement.
- No one is going to judge you. We may take your considerations into account when we talk to other specialists on this topic. What we need is reliable and understandable introductory information, nothing more.
- Since that's the case, I'll give it a try.

The brain has evolved in the same evolutionary way as the other organs. In the brain there are ancient structures located directly above the spinal cord and possessing a significant degree of autonomy, and a number of structures that emerged at later stages of evolution of the vertebrate branch at the top of which we are located.

Autonomy is ensured by a number of groups of neurons localised in the medulla oblongata, which are actually centres processing incoming internal information in the form of nerve impulses. The main part of impulses coming through ascending nerve channels contain some information about vital parameters of internal organs. Another part of impulses comes from the upper, later structures of the brain. Based on the results of processing this information, neurons send impulses that correct the functional activity of organs.

The first conclusion from the above is that the autonomy of regulatory centres of the medulla oblongata is relative. Thanks to the connections between this part of the brain and its upper parts, our consciousness can modulate the work of the autonomic nervous system. But this possibility is greater for some functions (e.g. respiration) than for others (e.g.

haemodynamics). It should also be noted that from some organs, nerve conduits go not only to the medulla oblongata, but also higher up. These connections provide a structural basis for the reverse modulation of our psycho-emotional state.

- Is the brain composed solely of neurons?

- Well, not only that. Firstly, if only because the brain, like all other organs, must have a vascular network. Secondly, apart from neurons, there are star-like cells (astrocytes), small glial cells, the number of which is about 10 times greater than the number of neurons, as well as cells from which the myelin sheath of conducting nerve fibres is formed. Their function is not yet fully understood, but it is known that they contribute to the metabolism of neurons.

- Why is the brain surrounded by fluid?

- Yes, indeed, both the brain and the spinal cord are immersed in a special fluid called *"cerebrospinal fluid"*. It can flow from one cavity to the other through a small gap between the head and the spinal canal. This fluid dampens the shock loads on the brain tissue during sudden accelerations. In addition, all metabolic processes in the cells take place through the pericellular fluid.

The spinal cord is the main mediator through which the nerves that connect the structures of the brain with the structures of the body run. At the exit of the spinal cord there are small clusters of neurons - ganglia.

- Is all communication between the body and the brain through these ganglia and the spinal cord?

- Not all of it. There is a nerve (called the vagus nerve) that goes roundabout and innervates almost all internal organs.

- I've heard that there's some kind of intestinal brain in the stomach. What is it?

- Actually, neuronal clusters aren't just in the intestine. They are present in almost all internal organs. Thanks to these local clusters, there is some autonomous regulation of the organ's activity and the state of its blood vessels. It is just that in the intestine the nerve network is rather dense, which is connected with the evolutionarily important role that the digestive system has played and continues to play in the survival of animals.

- Okay, so far we've talked mostly about the anatomy of the brain. What can you do to help us understand how it works?

- You're stumping me again! I will not be able to simplify this most complex organ into some primitive units to explain their functioning with the scraps of knowledge I have. The only thing I can do is to draw your attention to the relationship between cell type and its external functions.

So far, I have focused your attention on the fact that the functions of multicellular formations derive from the properties of the constituent specialised cells. Let us apply the same method to analyse the peculiarities of the brain.

Regardless of the particular type of neuron, it has a soma (body), one long branch (axon) and many short branches - dendrites. It is true that some receptor neurons are bipolar, that is, they possess two axons, one of which is associated with sensitive terminal branches, and the other - centripetal - contacts other neurons. All terminal branches of a neuron terminate in a thickening *synaptic plaque* (see Fig. 4).

There is no mechanical contact between the two neurons. The contact is functional and is carried out through the synaptic cleft, into which bubbles containing chemicals of different composition are released from the synaptic plaque under the influence of an impulse. They fulfil the role of stimulatory or inhibitory mediators. Through diffusion, these mediators reach the postsynaptic membrane (it can be either on the soma of another neuron or belong to one of its branches) and excite an action potential in the neuron. It spreads through the network of neurons and the process resumes. This is how internal information spreads in the network.

The presence of a synaptic cleft and chemical mediators of neuronal integration into a network indicates that the neuron is a later version of a normal cell that previously had only one way of interacting - through chemical mediators.

The synaptic cleft eliminates the propagation of the electrical impulse in the opposite direction.

This brief introduction was necessary in order to try to understand, at least at a conceptual level, what thought and intellectual activity are.

An individual neuron is only capable of producing an output impulse each time the amplitude and time summation of all incoming impulses is sufficient for an action potential - impulse - to arise in a special area of the soma (*axon hill*), where membrane excitability is increased.

Thought is always the result of ensemble function of neurons. Functional ensembles of neurons may include from hundreds of thousands to tens of millions of neurons. It seems that in such a network impulses circulate for some time without fading. There is an opinion that this circulation is the material basis of thoughts. They arise only under favourable conditions, when the impulses coming through all channels are almost synchronised. Since this synchronisation is almost accidental, it seems to us that thought is spontaneous and fleeting.

- How then can we organise our train of thought, reason logically?
- I warned you, the brain is a dark substance. We don't know the answer to your question.
But we know that thought has many modalities. When we have a thought about food, we remember the sight of it, the smell, the people we shared the meal with, the weather while eating it, conversations, anecdotes, and more. This multifaceted nature of thought suggests that a given neural network has connections to many other networks. In the language of the mathematician, thought is a function of many parameters. Certainly, among these parameters, our sensations, which have entered the brain through the senses, are at the top of the list. But there are also parameters that are the result of processing this primary information. That is why every word and every thought has a huge number of associations.
- Well, can you even tell me how autonomous the brain is?
- I think we have touched on this topic before. My subjective opinion is that the brain initially needs some minimal information load to work. Further, if there is no new inflow of information, it can work autonomously, figuratively speaking "play" with this information, create new information from it and give it out in the form we call knowledge. If the information is replenished, the brain compares it with the existing base, checks the compliance of the new information with the existing knowledge model and, possibly, corrects the model.
- It all sounds convincing, but it's very speculative. What is the factual basis for it?
- I don't know how much you'll be satisfied with my answer, but I want to use one example. American psychologists conducted a survey to find out whether people's worldview correlates with prevailing dream themes. It turned out that Americans' main dream themes were associated with financial success, work and money, while the two predominant themes among the Mexicans surveyed were food and religion.
- I get the hint! What you think about during the day, you dream about at night. One brain model of knowledge excites and supports the work of another.
- Exactly!
- Isn't that too primitive? After all, it happens that suddenly a thought will shine clearly in your mind, linking a known concept with a new one! What is it?
- Neurophysiologists have found that neurons build up new dendrites all the time and at the same time destroy some of the existing ones. The process of

growth of new outgrowths is slow. A new branch wanders in space and stops growing only when it encounters an obstacle. It can be either the soma of another neuron or its outgrowths. Immediately after mechanical contact, a new synaptic channel of information transmission begins to form at this place.

- You mean to say that a thought circulating in one ensemble of neurons for a while can flow into another, right?

- Figuratively speaking, this is what happens. This is the material, neural basis of spontaneously arising mental associations.

- And what accounts for the variety of associations evoked by odours?

- This is both a simple and complex question. Until recently, science was dominated by the view that each odour (and its carrier is a specific chemical molecule) has its own receptor in the nose. The molecule and its receptor were thought to fit together like a key to a lock. But the place where odour information is stored in the brain is an ancient structure - the limbic system. It is located above the brain stem. The limbic system receives information not only from the olfactory nerve, but also from the organs of vision, hearing, balance and equilibrium, and muscle proprioceptors. Some of the information processed by the limbic system is sent to the cerebral cortex and the reticular system, which activates all neurons in the brain. The limbic system, which represents the main centre of emotion, is closely connected to the thalamus, hypothalamus and pituitary gland. Uncovering this network organisation has formed the basis of our understanding of why smells trigger such a vast array of memories.

It has recently been shown that the key-lock principle itself is not quite applicable to explain our perception of odours. It turns out that there are quantum processes involved that are not yet quite clear. At least it has been established that it is not the entire "odour" molecule that is important, but its specific active component. In other words, different complex molecules that have a common fragment in their composition will cause the same response in our brain.

Why such an interest in smells?

- I've been thinking about how the brain's perceptual systems work. When you talked about sensors, you left out odour sensing. So I wanted to get some more information.

- Note that the brain's olfactory centre can change its size. And this is related to learning.

- What do learning and adaptation have in common?

- Nothing!

Adaptation is an intracellular structural rearrangement that preserves the structural integrity of a cell. We also know that adaptation of an organ is a change in the number of its cells, the formation of a new network of small vessels. Nothing of this kind is observed during learning.

When we learn to speak, read, memorise the multiplication table, the way home and master other similar skills, it doesn't occur to us to say that we are adapting to, say, the multiplication table!

- I once watched a programme on the Discovery Channel about London taxi drivers. The programme claimed something amazing: as they memorised a map of city roads (there are about 25 thousand of them in London), the brain part of the trainees' brains increased decently. What do you think about that?

- Yes, I'm familiar with this research. We are talking about increasing the size of a special organ of the brain, the *hippocampus*, which is located in the central region of the brain. The hippocampus is responsible for memory, and spatial memory. The enlargement of the hippocampus was the result of an increase in the number of connections - nerve fibres - between neurons.

I can add something from my own experience. Every time I started writing a monograph, after a while I began to be bothered by frequent headaches. At first I didn't pay much attention to it. They were just pains. I've had them before.

But when it started to recur with each new book, I wondered why.

It is known that there are no pain receptors in the brain itself. They are in the blood vessels and inside the elastic dura mater, which is located between the bone and the brain. Since the space between the brain and the dura mater is small and filled with fluid, an increase in blood flow or volume of the brain itself increases the pressure of this fluid, which is perceived by the dura mater receptors as pain.

There is a direct correlation between mental activity and increased blood flow to the brain. But it is unlikely that I was thinking intensely 24 hours a day. It's more likely that the increase in cerebral blood flow can only explain a short-term headache. Chronic pain could be caused by an increase in cerebrospinal fluid pressure due to increased brain volume. Considering that the period of preparation for writing each book lasted approximately two years, and during this time it was necessary to read and memorise a huge amount of diverse information, there are reasons to believe that significantly increased interneuron connections increased the total volume of the brain. This is also indicated by the fact that a couple of months after the completion of this particular strenuous intellectual activity, the pain disappeared.

- I'm familiar with that kind of pain. But I thought it was a migraine, so I took analgin. While we're on the subject of headaches, can you clarify two more questions? Why does a headache hurt at all if there are no pain receptors in the brain? What is headache associated with sudden changes in atmospheric pressure?

- In order not to give you a false idea about the exclusively biomechanical nature of headache, I would like to point out that there are headaches of chemical nature. A number of endogenous products of cell metabolism cause cerebral vasospasm. Prolonged deterioration of blood supply to the smooth muscles of these vessels is the cause of headaches of this kind. And the same mechanism is the basis of nagging prolonged pain in other organs.

I have already mentioned that increased blood flow causes headaches. Now let's look at the mechanism of increased blood pressure. Let's imagine the effect of increased atmospheric pressure on the whole body. The brain and its surroundings are in a rigid cranial box, which does not transmit external changes in air pressure inwards. And the rest of the torso is elastic. In the tissues, the pressure is approximately equal to atmospheric pressure, because the skin and muscles transmit external pressure inside the body almost without distortion. And inside the body are blood vessels with incompressible blood. Therefore, as external pressure increases, some blood will move towards the head, increasing the pressure and volume of the brain vessels. Stretched blood vessels are one source of headaches. Another source is the overflow of the venous sinuses of the brain and increased pressure on the mechanoreceptors of the inner lining. The overstretching of these receptors causes the sensation of headache.

- Why does my headache also occur when the atmospheric pressure drops? I am a frequent flyer and I have noticed that I often get headaches on aeroplanes.

- Let's first talk about the mechanism of the effect of atmospheric pressure jumps on the body, and then let's get back to the aircraft.

According to biomechanics, when the external pressure (atmospheric pressure) drops, the elastic blood vessels in the torso slightly dilate and the blood pressure drops. This is enough to increase the outflow of blood through the veins of the head. In the shrunken brain sinuses, receptors in the brain cause pain sensations.

Now about why a passenger of a modern airliner may experience a headache. The point is that, in accordance with safety requirements, aeroplanes are equipped with a cabin and cabin pressure machine. After gaining a certain altitude and then throughout the flight, this device sets the pressure inside the aircraft at 230 mmHg less than the normal pressure at sea level. The cabin pressure value of 530 mmHg, which corresponds to the level of normal pressure at an altitude of 2450 metres, is the cause of headaches. Note that the body reacts to such low pressure by trying to get rid of "extra" fluid. A long flight has a diuretic effect.

- I have two comment questions. Why is this value of air pressure in the aircraft chosen if passengers suffer from it? Does it mean that people who live permanently at altitudes of 2450 metres and higher should have headaches all the time?

- The answer to the first question is as follows: physiological mechanisms of the majority of healthy people compensate biophysical unfavourable phenomena by a complex of adaptive reactions without any serious ailment.

To answer the second question, I will have to remind you that atmospheric pressure also affects the partial pressure of oxygen in the inhaled air. Consequently, the cells will receive less oxygen and the rate of ATP synthesis will decrease. We have analysed the complex response to hypoxia in detail in our talks on energy and system physiology. There we also analysed the mechanism of adaptation to chronic hypoxia.

- That's it, you don't need to elaborate any further. I remember. So, the blood of highland aborigines has more red blood cells, lungs are larger, cells have more mitochondria, and there is a developed network of microscopic vessels.

Why do headaches also occur with geomagnetic disturbances? What is the relationship between solar activity and headaches?

- Science does not know the exact answer to the first question. There is a hypothesis that changes in magnetic field strength increase blood pressure. This hypothesis is justified by two independent facts. First: the resistance of the vessel to the flow of blood increases with the increase in its viscosity. The coarser the blood formations, the higher its viscosity. Second: red blood cells contain the protein haemoglobin, which contains iron. It is sensitive to magnetic field strength. On this basis, it is concluded that when the magnetic field strength increases, individual red blood cells stick together and form clots that impede the flow of blood through small arteries. As if there's a logical connection. But it is not quite clear to me why blood pressure should rise in response not only to an increase in geomagnetic field strength, but also to its weakening. Besides, the geomagnetic field strength itself is much weaker than those man-made magnetic fields in which modern urbanised man is immersed.

Now about the connection between solar storms and their terrestrial echo (according to the terminology proposed by Chizhevsky at the beginning of the last century). We know that we are protected from moderate fluctuations of the Sun's electromagnetic field by the geomagnetic field. It is another matter when perturbations on the Sun's surface are so strong that there is a large coronal mass ejection. I note that even in this case, there are two different mechanisms for the influence of this solar event - the breaking of the loop of the magnetic field on the surface of the Sun - on us. The first, fast mechanism, is caused by the headache-inducing electromagnetic pulse arising at the moment of the loop rupture. It is fast because it propagates at the speed of light and reaches the Earth in about 8 minutes, perturbing the geomagnetic field. The second, slow mechanism, is due to the physical movement of ionised particles of the Sun in space. The speed of this movement is significantly less than the speed of light. Usually these particles reach the Earth's ionosphere and perturb its magnetic field three days after the solar event.

It turns out that if we invent clothes with embedded conductors (e.g. made of nanomaterials), we can protect vulnerable people from geomagnetic disturbances!

- That's perfect. I think it will work regardless of what mechanism causes pathological human sensitivity to geo- and heliomagnetic disturbances.

Now let us return to the variability of brain mass. I told you, after all, that the number of nerve connections increases when we repeatedly repeat new information in an effort to memorise it. But inter-neuronal connections are also eliminated if the brain doesn't use them.

- So in order for something not to be forgotten, you have to read it out of your memory more or less regularly, or what?

- In principle, yes! But there is one important detail that also works in technical memory devices. Every time information is read, it is written again!

- And what happens if a third event is wedged in between these two events?

- Funny thing is, you're not the first person to have this idea. Psychologists are already doing this to rid patients of traumatic memories. The doctor under hypnosis induces these memories in the patient, combining them with new, more pleasant information. They say it works.

- And how quickly do new interneuronal synapses grow or are destroyed?

- I read somewhere that it takes about a couple of months for new associations to form. As for breaking them down, there is indirect evidence that it happens faster. For example, a fortnight's holiday at the beach is enough to cause IQ to drop by 5-10 points.

- I just had a crazy question, not exactly on our topic. Can I ask it?

- Come on!

- To what extent does a person's brain shrink with age and is there a loss of

mental capacity associated with this?
- It is believed that the maximum number of neurons is created by the end of the intrauterine period. Further on, only new inter-neuronal connections are formed. They reach their maximum by about 20 years of age, further the number of neurons decreases. It is not for nothing that it is said that the brain of a second-year student is much more powerful than the brain of his professors.
As for intellectual losses with age, they are more often caused by brain fragmentation rather than by a decrease in the number of neurons. In case of senile fragmentation, associative connections suffer, it becomes harder to concentrate attention, it is difficult to realise several modalities of information of the external world at the same time. Often, when trying to concentrate attention, side effects occur: a person starts crying, other emotional and vegetative manifestations appear.
- Look, that's enough! I don't want to live much longer. Is there anything good about getting old?
- If it's any consolation, there is. The space between the brain itself and the skull box increases with age. If young people in brain haemorrhages in almost 100% of cases die, because the functions of vital brain structures are disturbed, then older people in most cases of haemorrhagic stroke survive. And often there is even complete rehabilitation and restoration of temporarily lost functions.
- That's comforting! It's funny, but it turns out that it's generally better for your health to have a small brain. It's harder to live a smart life!
- Brain size and intelligence are not strictly correlated. Einstein's brain mass (though he was already 76 years old at the time) was 1255 g, which is 200 g less than the average male brain mass.
- Another stray thought. If I don't say it, I'll forget it. What about dreams like that, when a healthy person dreams of some disease and after a while he really gets sick with it?
- There are two possible explanations. The first: the disease has already begun, and at the initial stages some chemicals get into the blood from the diseased cells, which pass through the barrier between the blood and the brain and initiate corresponding dreams. The second: a person is imaginative, and his consciousness and subconsciousness reconfigure the brain centres of the medulla oblongata so that those, suppressing the activity of autonomic systems, disturb the normal activity of cells of specific organs.
Suffice it to say that chronic weakening of the regional blood supply can lead to organ pathology.

- What about your adaptive coping mechanisms?
- They are, of course, activated. But the possibilities of physiological mechanisms are not unlimited!
- I'll ask this curiosity question - which brain is more powerful, the one with more grey or white matter?
- The grey matter is the neurons of the cerebral cortex and the white matter is the nerve fibres between the neurons. Before answering, I would like to set one limitation: with the same mass of both brains.
Studies with the help of magnetic resonance tomographs have noted one curious detail: in representatives of exact sciences, first of all, mathematicians, the mass of white matter is significantly less than in people of humanitarian professions. And grey matter is often not so much. This puzzled the researchers. After all, it was believed that mathematicians had more developed brains! So what is it?
To find the key to this riddle, kindergarten-age children were studied. Two groups were identified among them. The first group included those who practically did not cheat or lie. The second included those who were prone to lying. The results of the CT scan surprised the researchers. It turned out that honest children have much less grey matter than children prone to trickery. Moreover, according to standardised tests, the intelligence scores of the "honest" ones exceeded those of the "tricksters". The researchers' conclusion was unambiguous: additional connections between neurons is a way to compensate for the lack of natural abilities.
Australian sociologists and psychologists conducted another long-term study. It showed that the lack of early social adaptation of poorly gifted children shortens their life expectancy. In other words, the presence of natural giftedness in any field contributes to the development of this specialisation, but to the detriment of socialisation. On the other hand, the absence of early signs of talent is not a social verdict. On the contrary, good adult mathematicians, for example, often work for those who did not achieve much success in their school years, but had to learn to lie to get ahead in society. A person who has many answers to the same question has a better chance of social success than someone who perfectly learns maths, physics and exact sciences. This is a paradox, but a fact!
- When I insisted on this conversation, I did not expect such a judgement! It turns out that you and I have learnt and done science for nothing.
- I think, dear, that we went into science because our parents and teachers praised us for our good studies. But our brains were socially underdeveloped, not building the right number of interneuron connections. We loved truth, not

money! And even if we had money, we would have spent it on setting up research labs, supporting science, wouldn't we?
- I guess you're right. There's something wrong with our genes.
- My dear, it is genetics that states that there must be such genomes in the population. After all, if there were no people of science, there would be no modern civilisation! So, let's end our talks on this optimistic note.
- Sorry, but you didn't say anything about emotions. I've encountered a lot of contradictory things in discussions on artificial intelligence regarding the need to model this aspect of human life. On the other hand, there are already humanoid robots with outward manifestations of certain emotions. Can you tell me your opinion?
- Emotions are a separate huge biological problem. I can hardly cover all its aspects. I will only note that emotions are evolutionarily fixed because they are used to solve important tasks of survival. For the average person, emotions are experiences of joy, fear, anger, longing, and sadness. But behind all the varieties of emotions there is an underlying physiology and biochemistry. For example, consider the physiological mechanisms and manifestations of fear.
There is a special paired organ in the brain (the *amygdala).* When the brain identifies information from different sensory channels as a threat, the amygdalae send impulses to the secretory glands of the adrenal glands. These glands increase the secretion into the bloodstream of two important hormones, *adrenaline* and *cortisol.* I think everyone has heard of adrenaline. Cortisol, on the other hand.
is more associated with the single-rooted therapeutic drug hydrocortisone, which has anti-inflammatory properties. Both hormones have a complex action - they activate the cardiovascular system, respiration and increase attention. At the same time inhibits the digestive system, increases blood coagulation. The complex of mobilisation reactions is aimed at protecting the organism from possible damage. Another component of this complex of reactions - increased sweating - prepares the body for the removal of excess heat, inevitable in a fight or run. But prolonged activation of the adrenal glands has negative consequences typical of stress: chronically increase blood pressure and appetite, a person gains weight. In extreme situations, the adrenal glands are surgically removed. It is funny, but after that often a person loses the feeling of fear altogether, although the amygdalae signal danger.
-*I* realised that emotions are a serious and delicate matter.
Since you are finishing educating me about life, I can't help but ask the most

important question, almost Schrödingerian. You claim that specialised human cells compete for food, and that the relationship between nerve cells and somatic effector cells is frankly antagonistic. How, then, did the multicellular organism as a "project" prove to be evolutionarily successful? What is the synergy of specialised cells?

- Yes, Erwin Schrödinger would commend you for this profound philosophical question. But to answer it, I'll start at a distance. Humans and animals represent only one group of multicellular organisms. Multicellularity first appeared as simpler organisms-colonies living in aquatic environments.

The exact evolution of these life forms into plant and animal organisms has not yet been unravelled by science. But an example such as a tree will help us understand the synergy of specialised cells. Leaves assimilate energy from the sun and store it in the form of sugars. Roots and branches participate in this process of photosynthesis: roots take out water and dissolved minerals, which are transported along the trunk and branches to the leaves via tubules. Without at least one link in this chain, the plant would not live. The fruits of the plant, containing the seeds that are spread by intermediaries, increased the plant's evolutionary chances in the struggle for a place in the sun.

Returning to the animal organism, some analogy with the synergy of cells in plants can be seen between the digestive system and those cells that contribute to the extraction of nutrient resources. The brain, which organises this process, contributes to this synergy. The synergy is enhanced by the fact that the brain also contributes to escape from predators. I think these two aspects of brain activity quite outweigh its oppressive aspirations to suppress the metabolism of effector cells.

- It turns out that, from the point of view of evolutionary mechanisms, intellectual activity unrelated to the survival of the human species is superfluous!

- As unfortunate as this conclusion may sound to us, our intellectual successes in the fields of culture and science are only a by-product of the evolution of the current dominant life form. But there is one essential detail. The conditions on our planet, and perhaps in the solar system, will not always be favourable to life. Only this by-product of evolution has the potential to preserve life of its own and other species by moving them to more suitable planets in space.

CHAPTER 13

Conversation thirteen

"Exam."

- As we agreed, you will be the presenter today. We need to see if you have learnt the rules of cells and understood the principles of the human body. Also today's friendly examination should show whether you have understood the main causes of maladaptation diseases that often accompany aging. Let your answer to these questions be the conclusion of our discussions. Agreed?
- I agree. I think I'm ready.
- And I am ready to listen patiently. With your permission, I will make short remarks along the way, if necessary. Let's go!
- There is no single definition of life yet. Biologists only agree that there is no life without cells. Each cell is a complex biochemical machine. It has specialised structures - organelles, with the help of which it is able to synthesise its own macromolecules to support life. For biosynthesis, the cell needs an influx of substrates and energy sources. Biological macromolecules are sensitive to changes in physicochemical conditions and under their influence spontaneously degrade. It is only because the rate of biosynthesis slightly exceeds the rate of this decay that the cell continues to live. The individual cell is mortal. But life as a phenomenon persists because the cell has time to reproduce itself by division. The time from one division to the next is called the cell cycle. In different phases of the cell cycle, the rate of biosynthesis is not the same and depends on the influx of substrates.

Waste products of cell activity seep through the pores of the cell membrane into the pericellular space (usually liquid). The cell membrane is one of the main structures of the cell, thanks to which a physicochemical environment is created and maintained in the cytoplasm that differs from the environment of the cellular fluid. It is due to this difference, maintained by ion pumps, that the cell is able to respond to external changes and carry out optimal life activity (metabolism). Two preconditions are important for normal metabolism: timely supply of nutrients and timely removal of metabolic wastes. These conditions are important not only for a unicellular organism, but also for the cells of all multicellular organisms. Moreover, it is precisely to solve these basic cellular problems in multicellular organisms that evolution has fixed specialised organs and their functional systems.

The terrestrial life so far familiar to us is considered a unique phenomenon. There are two kinds of life: unicellular and multicellular. Although both have a huge variety of variations, all life forms are based on a common way of encoding hereditary information using a four-digit code. It is based on four molecules: adenine, guanine, thymine, and cytosine. In fact, these molecules are the "letters" that make up the "words" of life - genes. Each gene has its own length, which is associated with many difficulties in deciphering the genome of humans or other organisms. The genes form DNA, which is located in the nucleus of the cell. The DNA double helix consists of strands of RNA molecules joined by adenine, guanine, thymine and cytosine. The basis of DNA replication during each cell division is that adenine of one strand of RNA corresponds to thymine and guanine to cytosine on the other strand.

DNA is not a whole macromolecule, it is fragmented into many parts - chromosomes. Each organism has its own unique set of chromosomes. It is known that in each chromosome only a small part of genes is active, i.e. takes part in protein synthesis. The remaining genes are called "dormant" or "silent".

The reliability of DNA copying is not absolute, so copying errors often occur. Random changes in the structure of DNA that are passed on to daughter cells are called mutations. It was thought that it was mutations that created all the diversity of life from the original mother cell. However, it has now been established that besides mutations, there is another - epigenetic - mechanism for passing on hereditary information to offspring. It can deactivate active genes or "wake up" dormant ones and make them encode proteins. Specialists already know how to operate not only individual genes, but also the epigenetic mechanism for targeted changes in inherited properties.

Life has been evolving for almost four billion years. For more than three-quarters of that time, life has been in the form of a variety of single-celled organisms. During the first billion years, the cell had no nucleus, the way in which hereditary information was passed on to daughter cells was unreliable, and frequent mutations led to the emergence of new cell types.

Initially, oxygen was scarce in the planet's atmosphere and was poison to anaerobic microorganisms. At some point in the initial phase of intense evolution, cells appeared among the anaerobic cells that used the energy of sunlight to photosynthesise hydrocarbons. The glucose produced from the hydrocarbons underwent anaerobic glycolysis to form two ATP molecules. One of the by-products of the metabolism of these cells was oxygen, which gradually accumulated in the atmosphere. At this stage of evolution, mutations gave rise to a new form of life - a small, beautiful microbe. And it was beautiful because it had the ability to synthesise ATP from carbohydrates and oxygen. This is how aerobic ATP synthesis came into being, the

efficiency of which is almost 17 times higher than that of anaerobic glycolysis.
Then, in the course of the more or less uniform evolution of life, another remarkable event happened. It consisted in the fact that one of the large cells with a nucleus "swallowed" this wonderful microbe of ours, which is able to extract benefits from oxygen. Moreover, in the process of absorption the cell did not digest the microbe, but took a part of its genetic information to itself in the nuclear DNA, and the other, small part, about a hundred genes, remained in this organelle-microbe. It turned out to be a kind of symbiosis between the two organisms. This organelle is known as the mitochondrion, often referred to as the energy substation of the aerobic cell. The number of mitochondria in a cell is variable: mitochondria are able to multiply and do so when there is a shortage of ATP molecules in the cell. The greater the number of such ATP producers, the higher the host cell's energy capacity and its efficiency. Thus, the energy prerequisite for the formation of a new form of life - multicellular organisms.
Biologists still don't know how this metamorphosis occurred. There are several hypotheses. What is important for us is another: the new kind of life had to solve all the major problems that unicellular life had encountered and overcome. In other words, a multicellular organism had to get food, assimilate it, distribute it to all its constituent cells, respond to external threats, and somehow reproduce itself. And all this against the background that all this set of functions remained with each of its constituent cells. It's not an easy task. And it's been solved in many ways. At least, the right to such a statement gives us a huge variety of plant and animal forms of multicellular organisms. Although each of them has specific "know-how", we have much in common with them. It is this thesis that gives physiologists and biochemists the right to base their studies of the human organism on analogies with other multicellular organisms.
- I have to say that you have quite dashingly and imaginatively laid out not only what I've been talking about in our conversations, but also clearly used additional information. Bravo!
- But you don't have to shout "Encore!" I'll continue.
Now about the human body. It contains trillions of cells belonging to one of 220 cell types. This variety of cell types is derived from a fertilised egg during the intrauterine process of individual development - ontogenesis. A cell type or specialisation is its external function, the users of which are cells of other specialisations. Often what we call the external function or purpose of a cell of the body is a by-product of its metabolism or the result of a

response to an external destructive influence.
The main form of existence of cells of each specialisation is a colony (population) of sister cells. Moreover, in a colony, cells of the same type are rarely absolutely identical. Their microscopic differences determine the peculiarities of each cell's response to a common external influence. As a consequence, the functioning of the whole colony is slightly different from that of a single cell. This property of population response to an external signal has played a decisive role in the organisation of sensory organs. Even without increasing the accuracy of an individual sensor, the organ-population successfully transforms the continuous dynamics of the input environmental parameter into patterns of impulses sent to the brain.
Within each of our cells, biochemical circuits convert input chemicals into chemicals that become energy sources (ATP) or building blocks for the construction and repair of cell organelles. Repair is necessary because biological macromolecules are unstable and often disintegrate.
Most biosynthetic processes require energy. It is released during the breakdown of phosphorus bonds of macroenergy molecules (AMP, ADP and ATP). The greatest amount of energy is released during the breakdown of ATP. Glucose, other carbohydrates, fatty acids and fats are the primary consumables from which the mentioned universal macroergues are synthesised in two ways (anaerobically - in cytoplasm and aerobically - in mitochondria). Mitochondria, whose number in the cell is proportional to the average rate of energy expenditure, are the main suppliers of ATP. It is important to emphasise that the ionic homeostasis of the cytoplasm is so critical for body cells that in the process of evolution of multicellular organisms special organs and their functional systems have been selected to help cells maintain this homeostasis.
Each cell also has intracellular mechanisms to combat ATP deficiency. The best known is the chemical negative feedback mechanism that uses the ratio of ADP/ATP concentrations in mitochondria. As this ratio increases, the rate of ATP synthesis increases. This rapid mechanism optimises the rate of ATP synthesis when ADP and ATP concentrations fluctuate frequently. Mitochondria themselves are mobile and sensitive to oxygen concentrations in the cytoplasm. Mitochondria accumulate in areas where there is more oxygen. This is especially important in nerve cells with long branches - dendrites and axon. It is at synapses that a significant portion of ATP is consumed. There are other intracellular regulators of ATP synthesis, but the greatest role in the fight against chronic energy shortage is played by the total area of mitochondria of the cell. It is for this reason that the size and number

of mitochondria begin to increase in cells under prolonged stress. This, for example, occurs in the remaining kidney after surgical removal of its pair, in hepatocytes during partial liver resection, in muscle cells of athletes. Thus the organism makes full use of this intracellular mechanism, as there is no need for intensification of either blood circulation or respiration. Such prolonged intensification is energy-consuming and too wasteful for the body. However, cellular problems often involve too many cells. To solve such problems requires activation of the whole organism or those of its organs that are directly or indirectly involved in the creation of a favourable cytoplasmic environment.

There is ample reason to believe that our anatomy and physiology are the way they are largely because our cells needed it. Figuratively speaking, our cells created us! This idea serves as a key to understanding the workings of the body's individual physiological homeostatic systems and their interaction with the organism's mechanisms of adaptation to established physical and chemical environmental conditions. This interaction explains the emergence of causal chains producing unique physiological functions that are not peculiar to any cell. Examples of such functions are the contractile act of muscles, control of the secretion of specialised chemicals serving as secretions of the digestive and other systems.

Continuous life support of each cell of the body is a necessary condition for its physical and functional integrity. All kinds of intellectual activity, neuromuscular activity, as well as neurohumoral integration of various organs to achieve functional integrity and active existence of the organism would be impossible without excitable cells. In the process of our evolution, specialised organs and functional systems were formed that more or less successfully solve the tasks of life support of each cell. The leading role in this case belongs to three complex systems: digestion, respiration and blood circulation.

The blood is that complex fluid which consists of water, form elements (erythrocytes, leucocytes, etc.), substances which have leaked from the digestive system, and metabolic products of all the cells of the body. A large proportion of these metabolites have an influence on processes in cells of other specialisations. Some substances are inhibitors of biochemical transformations, others are stimulants. Due to such properties, blood mediates almost all physiological reactions of humoral type. Biochemists and physiologists have discovered only a part of physiologically active endogenous agents, so modern understanding of the regularities of integrative human physiology is still far from understanding the true causes of many pathological shifts. Endocrinology has established only the main pathways of such integration, but its nuances are yet to be discovered.

The site of substance exchange between blood and cells is the smallest vessels - precapillaries, capillaries and postcapillaries. These are porous vessels from which blood gases and water with small chemicals dissolved in it leave and enter by the laws of diffusion. The process of diffusion occurs because of the difference in the partial pressures of the gases, because of the difference in the concentrations of the substances, and because of the difference in the pressures of the fluid. The value of capillary blood pressure is the base that determines the value of mean arterial pressure, taking into account its fall to overcome the resistance of arteries and arterioles. Hence the negative feedback between the cells not satisfied with the inflow of necessary nutrients and oxygen, on the one hand, and blood pressure, on the other. In other words, by increasing blood pressure, the body improves cell metabolism. But this is only one of the mechanisms for optimising the vital activity of cells. In case of ATP deficiency caused by hypoxia, all mechanisms of increasing oxygen concentration in the incoming blood solve this problem without increasing blood pressure. In other words, intensification of lung ventilation, increase of their geometric dimensions (in chronic cases of hypoxia), growth of blood erythrocyte number lead to the general effect - elimination of hypoxic hypoeurgy.

In cases of hypoergy due to carbohydrate deficiency, another multistage mechanism is activated, aimed at intensification of AMP synthesis.

The special role of blood as a carrier of everything that is necessary for cells or interferes with their optimal vital activity has made the cardiovascular system a participant in almost all mechanisms of life support. Evolution has selected a number of nervous and humoral mechanisms that regulate blood pressure. The biophysical basis of such regulation is pressure sensitivity to changes in certain parameters of the cardiovascular system: pumping function of the heart, hydraulic resistance and vascular tone, and blood volume. Any regulation of blood pressure is reduced to decrease or increase the current value of these parameters.

The current values of the parameters have developed in such a way that the long-term average needs of the body cells are optimally satisfied without activation of the brain. If the consumers of circulatory function steadily increase or decrease their demands in blood flow, the adaptation mechanism slowly brings the parameters of the cardiovascular system to new optimal values.

The optimal coexistence of cells in a single organism is to minimise the number of cells that have metabolic problems. When such problems occur, due to lack of energy and/or cytoplasmic contamination, some of the biochemical transformations in cells are interrupted. The intermediate chemical agents remaining in the cell or released into the lymph and blood so rearrange the modes of functioning of the body organs that restoration of optimal cell metabolism is ensured in the shortest possible time.

Both external and internal factors can be disruptors of optimal metabolism. The main internal disruptor of cell rest is the brain. By sending stimulating or inhibitory impulses to organs to organise their combined functioning, the brain disrupts the natural course of metabolism of the cells of these organs. This mode of functioning is destructive for cells, so it cannot last. Thanks to ancient intracellular mechanisms, the cells soon stop obeying the brain's commands and enter the recovery mode.

From what I understand, much in the body is geared towards building the kind of structure that provides for the needs of the cells while minimising energy expenditure. An example is the constant process of forming new microscopic vessels under chronic regional energy shortages. Because evolution has preserved many ways to achieve a common goal function, optimisation is always multi-parametric.

- I will note that such adaptive self-adjustment of each organism to chronic conditions of its existence automatically increases the contribution of those mechanisms that give the greatest effect. But I will also remind that in the process of ontogenesis each individual forms his own individually optimal combination of these mechanisms. That is why I insist that it is impossible to assess the health of all people using a common test. Each person has his or her own individual parametric picture of health. Although this picture may change at different times of life, a person remains healthy !

Thus, there are no major and minor cells in the community of our cells. Human health is the optimal coexistence of its cells. Any deviations from this rule of cells give rise to pathology.

Mitochondria, which gave energy to the aerobic cell, are too often involved in the defence reactions of the organism. Small size of mitochondria genome and frequent acts of its copying lead to age-related accumulation of mutations in mitochondria, energy deficit, because of which some cells of the organism die. The decreasing number of cells in any of their populations excludes the smoothness of distribution of cell anisotropies. The former smooth interaction between populations (organs) is transformed into a jump-like interaction. Frequent spikes in blood pressure and poor health are vivid manifestations of this mode in elderly people.

- Well. I think you've passed the exam. Thank you for listening and remembering. I hope that this acquired knowledge of man and his cells will be useful.
- I'm the one who should be thanking you for agreeing to talk to me about such a hot topic, which is health at our age.
Since life is like that, let's be healthy!

In lieu of an epilogue

From birth we are surrounded by people. At first it is parents and family. The circle of people with whom each of us enters into relationships expands over the years. This is how we are formed as members of a society that has its own rules: some relationships between people are encouraged, others are taboo. This is not only how human communities are organised, but also animal communities.

As an element of society, each member is a biosocial being. I suppose that before reading this book, you had not given much thought to your biological nature. When asked the simple question "How is life?", the first thing that came to your mind were events related to your social affairs. Historically, the social dominates over the biological in our minds. I don't think that thinking about your constituent cells was your first association with the title of the book "That's How Life Is". Now you have a new one - an inner vision of life. It will change you.

Usually, a society constructs rules for the coexistence of its members. The individual either accepts them or leaves. A multicellular organism is a special type of society: the constituent cells have no choice.

Often an animal community has a hierarchical structure with a dominant individual at the top. The history of mankind suggests that we are not far removed from the animal world. For a long time, humanity existed as small kin groups competing for territory and resources. Then larger ethnic groups with their own leaders emerged, but feuds and raids on neighbours continued even when states were formed on the basis of ethnic groups. Only the first empire in Mesopotamia led the peoples of that region to lasting peace and socio-cultural prosperity. The useful role of the new organisation of society is also confirmed by other isolated empires that emerged about five thousand years before us in the basins of the Nile, Ganges, Huang He and Yangtze rivers. However, even the imperial form of government was not devoid of human biological vices and inevitably gave rise to despotism, slavery and oppressed the personality of subordinates.

It is known that at different times and in different empires there were thinkers who proposed more just, in their view, forms of government. Two such forms are best known: Confucianism in the East and democracy in ancient Greece. However, neither Confucius (551-479 B.C.) nor the father of democracy, Plato and his disciple Aristotle (4th century B.C.) could even think that certain qualities inherent in man are conditioned by his internal structure. More than two thousand years later, science proved the cellular structure of organisms. But even after the establishment of this fact, the leading role of human cells in the formation of his biosocial personality is often ignored. Thus, the famous ideologists and politicians of communism believed that the essence of man could be radically changed through education. Indeed, inculcating a young member of society with new ideals and social values changes something in his personality. But not everything. In my opinion, this idea continues to be underestimated in the modern world. Instead of life as an unconditional universal value, local value systems dominate in states of different religious and political-legal systems. Although each such system is internally consistent, problems arise at the intersection of different systems. Moreover, military conflict is not always caused by a violated economic interest. There are a growing number of proponents of the view that fundamental changes in our biology are needed....

Science elucidates patterns in the reality around us. How scientific truths will be used is up to the users. Engineers need this knowledge to develop useful devices that make life easier, more interesting and comfortable. Doctors use scientific knowledge to develop medical technologies that reduce human suffering and lengthen life. However, in the hands of scoundrels, scientific knowledge, engineering and medical technology become a means of satisfying unhealthy animal instincts and dominance ambitions. As long as life continues to be like this, only enlightened public morality and humane laws will be able to put a stop to such animal aspirations.

I was recently invited to come to Germany and speak to a group of potential investors supporting the development of medical technologies and devices. They were interested in whether my understanding of the human body as a community of specialised cells could contribute to the birth of new translational and innovative medical technologies. The letter emphasised that the particular interest of the investors concerned three aspects:

1. Can sensory cells be modified to perceive additional physicochemical characteristics of the internal and/or external environment?

7. Is it possible to extract information about the energy status of each cell in order to create technologies for early diagnosis of cellular problems and their correction?

3. How should technologies for measuring venous blood chemistry be developed to detect early manifestations of organ pathologies?

I'm interested in the proposal. I am working on the report. Perhaps I will write more about the results of this meeting.

Scientific monographs by the author

The main monographs and publications in physiological journals in which the author has scientifically substantiated the provisions of this book:

1. Grigoryan R.D. Self-organisation of homeostasis and adaptation. Kiev (2004).ISBN 966-8002-99-7.
2. Grigoryan R.D. Biodynamics and models of energy stress. Kiev (2009). ISBN 978-966-02-5393-3.
3. Grygoryan R.D. The energy basis of reversible adaptation. N.Y., USA, (2012). ISBN 978-1-6208-093-4.
4. Grigoryan R.D., Lyabakh E.G. Arterial pressure: Rethinking. Kiev (2015). ISBN 978-966-02-7781-6.
5. Grigoryan R.D. The paradigm of "floating" arte-rial Pressures. Düsseldorf, Germany (2016). ISBN: 978-3-659-60433-1.
6. Grygoryan R.D. The optimal circulation: cells' contribution to arterial pressure. N.Y., USA, (2017). ISBN 978-1-53612-295-4.
7. Grygoryan R.D., Sagach V.F.. The concept of physiological supersystems: New stage of integrative physiology. International Journal of PhysiologyandPathophysiology,(2018): 9,2, 169-180.
8. Grygoryan R.D. The Optimal Coexistence of Cells: How Could Human Cells Create The Integrative Physiology. Journal of Human Physiology. (2019).1:8-25.
9. Grygoryan R.D. Milestones of the Modeling of Human Physiology. Journal ofHumanPhysiology. (2019).1:8-25.
10. Grygoryan R.D. Several Theoretical and Applied Problems of Human Extreme Physiology: Mathematical Modeling. Journal of Human Physiology. (2021).3,2:57-70.
11. Grygoryan R.D. Modeling of Mechanisms Providing the Overall Control of Human Circulation. Advances in Human Physiology Research. (2022), 4,1,.5-21.

Annotation

Grigoryan R.D. "Know thyself: synergy and antagonism of cells". 2024,- 214 c.

Popular science monograph is devoted to the physiology human beings. Through dialogues with an artificial intelligence specialist, a physiologist outlines the basic principles of the organism. For the first time, the organism is considered as a community of 220 types of selfish cells. All cells of the body (and there are about 50 trillion of them) compete for common resources. No cell "wants" to work for another cell, as this requires the expenditure of its limited energy. Nevertheless, the products of life activity of some cell types turned out to be useful for others. Thus, in the course of evolution, chains of synergy and feedbacks appeared. Such causal chains between different types of cells (e.g. neurons and muscle cells, neurons and secretory cells), as a result of which the organism escaped predators or obtained food, also turned out to be useful for the organism. Evolution has preserved only those species whose cells had the ability to counteract the "despotism" of the brain and restore optimal metabolism. Humans are healthy as long as biochemical and physiological multilevel mechanisms ensure a compromise between effector cells and the brain, the main organiser of behavioural response. It is shown how slowly developing diseases of age are associated with disturbances in this compromise.

Hopefully, mastering the knowledge of the fundamental rules of cellular coexistence will help us not to violate them and be healthy.

Printed by Books on Demand GmbH, Norderstedt / Germany